LUSCHIIM'S PLANTS

LUSCHIIM'S PLANTS

Traditional Indigenous Foods, Materials and Medicines

DR. LUSCHIIM ARVID CHARLIE
and
NANCY J. TURNER

HARBOUR PUBLISHING
www.harbourpublishing.com

HARBOUR PUBLISHING CO. LTD.

P.O. Box 219, Madeira Park, BC, VON 2H0

www.harbourpublishing.com

ALL PHOTOGRAPHS by Nancy Turner, except: page 35 (goldenback fern) by Jim Morefield, CC BY-SA 2.0: https://creativecommons.org/share-your-work/licensing-considerations/compatible-licenses; page 94 (arbutus bark) © danieljemartin / Adobe Stock; pages 96–97 (arbutus trees) © Christopher / Adobe Stock; page 106 (black cottonwood trunks) by Genevieve R. Singleton; page 109 (domesticated plum) © Maria Brzostowska / Adobe Stock; page 142 (tall Oregon grape) by Genevieve R. Singleton; page 198 (deltoid balsamroot in a meadow), page 199 (deltoid balsamroot blooms) and page 215 (yellow avalanche lily) by David Polster; page 252 (cattail) by Ryan Hodnett, CC BY-SA 4.0: https://creativecommons.org/licenses/by-sa/4.0/deed.en; and page 274 (Nancy Turner and Luschiim) by Robert D. Turner.

INDEXED by Nicola Goshulak

COVER AND TEXT DESIGN by Shed Simas / Onça Design

PRINTED AND BOUND in South Korea

Harbour Publishing acknowledges the support of the Canada Council for the Arts, the Government of Canada, and the Province of British Columbia through the BC Arts Council.

CATALOGUING DATA AVAILABLE FROM LIBRARY AND ARCHIVES CANADA

Title: Luschiim's plants : traditional Indigenous foods, materials and medicines / Dr. Luschiim Arvid Charlie and Nancy J. Turner.

Names: Charlie, Luschiim Arvid, author. | Turner, Nancy J., 1947- author.

Description: Includes bibliographical references and index.

Identifiers: Canadiana (print) 20210164964 | Canadiana (ebook) 20210167602 | ISBN 9781550179453 (softcover) | ISBN 9781550179460 (EPUB)

Subjects: LCSH: Ethnobotany—British Columbia—Pacific Coast. | LCSH: Plants, Useful—British Columbia—Pacific Coast. | LCSH: Plants, Useful—British Columbia—Pacific Coast—Identification. | LCSH: Plants—British Columbia—Pacific Coast. | LCSH: Plants—British Columbia—Pacific Coast—Identification. | LCSH: Indigenous peoples—British Columbia—Pacific Coast.

Classification: LCC QK98.4.C3 C43 2021 | DDC 581.6/3097111—dc23

CONTENTS

NOTE REGARDING PLANT USE, SAFETY AND CONSERVATION

———————

THE INFORMATION PRESENTED HERE, PASSED ON FROM THE TEACHINGS Luschiim has received from his own Elders, and from his personal experiences and observations, is intended for educational use. Many of the plants described here are not as common as they once were, so if you plan to harvest any plants, please ensure that you do so in a sustainable way, guided by how common and plentiful the plants are and the impacts that harvesting might have. Never collect plants in protected areas or on private lands without permission.

Also, some plants are potentially harmful, from the chemicals they contain, from thorns or prickles, or in the case of stinging nettle, from stinging hairs. Some plants can be confused with poisonous look-alikes. For example, the bulbs of death camas (*Zigadenus venenosus*) closely resemble those of the edible blue camas species (*Camassia* spp.). The former can be deadly, even if only one or two bulbs are consumed. Blue-flowered camas bulbs, on the other hand, have been an important food for Quw'utsun (Cowichan) and other Coast Salish peoples of Vancouver Island for thousands of years. When properly cooked, they are a sweet and nutritious food. These species—death camas and edible camas—often grow together in the same kind of habitat, so it is important to be able to tell them apart. Always take care to confirm the identification of plants you may wish to use and learn the proper timing and methods of harvesting and preparing them.

In particular, as with prescription medicines you might purchase from a drug store, many traditional plant medicines can be harmful if not properly prepared or if taken in the wrong dosage. There can be variable concentrations of medicinal ingredients contained in different parts of the same plant, at different seasons or within populations of the same species. Traditionally trained plant specialists understand the conditions under which medicinal plants are harvested and administered. Never take medicines without first checking with a qualified doctor or herbal medicine specialist who can give you advice regarding suitability, preparation and administration of these medicines.

Traditional medicine specialists believe that plants, like people and animals, have their own spirit or life force. Any plant you wish to harvest and use must be treated with great respect. Talking to the plant and asking for its help and permission to harvest it may seem strange to those coming from an urban, Western perspective, but this is standard practice for many Indigenous users of plants and other resources.

The conservation of these precious plants and their habitats is always of major concern, along with your safety. It is very important to look after these plants and the places where they grow so that they do not disappear. You might follow the lead of those practising ecological restoration at the Cowichan Garry Oak Preserve (run by the Nature Conservancy of Canada) and elsewhere in the Cowichan region. Some of these plants can be grown easily in your garden from seeds or cuttings, or even grown as attractive potted plants.

There are a number of plants mentioned here that have special ritual applications in Hul'q'umi'num' ceremonial contexts. The knowledge about these is private and has not been provided, other than an indication that such spiritual significance exists, but only those with the rights to such knowledge have access to it. On the other hand, cultural knowledge of the day-to-day uses of plants for foods, medicines or in technology is meant to be shared widely and passed on to future generations, for increased understanding and valuing of the plants and the places where they grow.

We recognize the private and sacred nature of medicinal plants and other preparations used ceremonially or ritually by the Quw'utsun people. Luschiim's aim in sharing knowledge about the identity of these plants is to ensure that they are remembered, and that their cultural importance is passed on to the future generations.

NOTE ON LINGUISTIC WRITING SYSTEM FOR HUL'Q'UMI'NUM' PLANT NAMES AND OTHER TERMS

THE WRITING SYSTEM WE USE HERE IS A PRACTICAL ONE USING ENGLISH alphabet letters, many of which sound as they would in English, while others represent sounds that are not necessarily found in English words. The *FirstVoices* website provides a key to the pronunciation of the sounds, briefly summarized as follows:

'	Glottal stop (sudden catch in the throat) preceding a vowel at the beginning of a word, or a glottal mark, following a consonant (including resonants, r, m, n, y, l), indicating glottalized or "exploded" sound
a	Pronounced as in English "father"
aa	Similar to "a" but held longer
ch	Pronounced as in English "cheap"
ch'	Pronounced like "ch" but with a slight popping or exploding sound; rare in Hul'q'umi'num'
e	Mostly pronounced as in English "bet" or "bat"
ee	Similar to "e" but held longer
h	Pronounced as in English "heat"
hw	Pronounced with the lips rounded, as if your mouth is ready to whistle; a soft "whooshing" sound
i	Similar to English "meet"
ii	Similar to "i" but held longer
k	Occurs in borrowed words; pronounced as in English "key"
kw	Pronounced as in English "queen"
kw'	Pronounced like "kw" but with a glottalized or popping sound
l	Pronounced as in English "long"
lh	Pronounced a little like "shoe" in English but with the tongue pressed up behind the front teeth so that air is released from the sides of your mouth

m	Pronounced as in English "**m**eet"
n	Pronounced as in English "**n**eat"
o	Pronounced as in English "**oh**"
o'	No English equivalent; used at end of words or syllables
oo	Pronounced as in English "m**oo**n" but held longer (long version of "ou")
ou	Pronounced as in English "h**oo**t" (in some Hul'q'umi'num' words borrowed from French, Chinook or English)
p	Pronounced as in English "**p**ond"
p'	Pimilar to "**p**" but glottalized, with an exploding or popping sound
q	Similar to English "**k**" but with the tongue further back in the throat
q'	Similar to "**q**" but glottalized, with an exploding or popping sound
qw	Similar to "**q**" but with lips rounded, as in English "**qu**een"
qw'	Similar to "**qw**" but glottalized, with an exploding or popping sound
s	Pronounced as in English "**s**it"
sh	Pronounced as in English "**sh**ore"
t	Pronounced as in English "**t**ake"
t'	Similar to "**t**" but glottalized, with an exploding or popping sound
th	Pronounced as in English "**th**in"
tl'	Made by holding the tongue as for a "**t**" sound, then releasing the sides, as in an "**l**" sound, with a clicking or exploding sound
ts	Pronounced as in English "ba**ts**"
ts'	Similar to "**ts**" but glottalized, with an exploding or popping sound
tth	Pronounced as a single sound combination of "**t**" and "**th**," as in "cut **th**in"; found only in a few Hul'q'umi'num' words
tth'	Similar to previous "**tth**" but glottalized, with an exploding or popping sound
u	Pronounced as in English "b**u**tter"
w	Pronounced as in English "**w**ill"
x	Pronounced with the back of the tongue near the soft palate—as if starting to clear the throat
xw	Similar to "**x**" but with lips rounded
y	Pronounced as in English "**y**ellow"

ACKNOWLEDGEMENTS

WE ARE GRATEFUL TO ALL THE HUL'Q'UMI'NUM' KNOWLEDGE HOLDERS OF PAST generations, whose knowledge and wisdom are reflected in Luschiim's teachings. We would also like to thank Luschiim's family; his wife, Darlene; their children and grandchildren; and his parents, Simon and Violet Charlie. Luschiim's cousin Al Scott Johnny of Nooksack has been a particularly good source in terms of corroborating Luschiim's knowledge from his own experiences with plants. Thanks to those linguists dedicated to the recording and documenting of Hul'q'umi'num' language, including Professors Tom Hukari and Donna Gerdts, Ruby Peter and many Elders in the Hul'q'umi'num'-speaking community going back to the early 1970s and before, whose work resulted in the Hul'q'umi'num' Dictionary, available online. Thanks to Cowichan Tribes FirstVoices Language Administrator Chuck Seymour and all those responsible for the fantastic work of *FirstVoices*. We would also like to thank Felix Jack, Tim Kulchyski, Dianne Hinkley, David Bosnich, Greg Sam, Genevieve Singleton and David Polster, Nejma Belarbi, Liz Hammond-Kaarremaa, Pamela Spalding, Trevor Lantz and Kate Proctor for participating in this work over the years. We are grateful to Nancy's husband, Robert D. Turner, for his photography and for helping with this work in so many ways. We hope that this precious knowledge will be appreciated and applied by members of the Cowichan Tribes and other First Nations of the region to help continue the work and responsibility of caring for the land and all the life that it supports.

INTRODUCTION

LUSCHIIM, DR. ARVID CHARLIE, IS A RESPECTED ELDER AND BOTANICAL EXPERT of Cowichan Tribes, and a fluent speaker of his Hul'q'umi'num' language. His knowledge of plants is truly remarkable and comes from deep training and experience, starting in his earliest childhood years. He learned this knowledge from his great-grandfather Luschiim (whose name he inherited), his great-grandmother and others of their generation, who grew up in the last decades of the nineteenth century. In July 1999 Nancy Turner had the opportunity to meet Luschiim for the first time, in a field workshop with Cowichan Tribes Treaty Office and Cowichan Community Land Trust. On this occasion, they walked with ecological stewardship trainees through dense Douglas-fir woods along the Cowichan River and were delighted to share their mutual love of the plants they encountered with a group of Quw'utsun youth. After a few more meetings, it was clear to Nancy that

Dr. Luschiim Arvid Charlie, on trail to Mt. Arrowsmith, 2010

Luschiim's knowledge was exceptional and that it was built on the wisdom and experience of generations before him. Furthermore, a born teacher, he wanted to ensure that the rich education he had received from his great-grandfather, grandparents, parents and others knowledgeable in cultural and environmental aspects of Quw'utsun life would be passed on to future generations.

We (Luschiim and Nancy) soon decided that it would be a worthwhile and important project to record Luschiim's botanical knowledge, a task that has been pleasant and meaningful for both of us. Our work on this project started in May 2005 and continued, with occasional interruptions, until 2020. We spent time out on the land, on Mount Prevost, near Mount Arrowsmith, at the Somenos Garry Oak Protected Area and elsewhere, as well as indoors talking about Luschiim's experiences with plants. Our major interview sessions are listed in the Sources section at the end of the book.

Luschiim was born in Quamichan, one of the villages of the Cowichan Nation, in 1942. His mother, Violet, passed away in December 2016. His father, the famous carver and artist Simon Charlie, passed away in May 2005 at the age of eighty-five. Luschiim, his namesake great-grandfather, who was born in 1870, lived until Arvid was about six years old and had a big influence on his life, teaching him about plants and medicines even at the tender young age of three to four years old. Even as a boy, Arvid was a hunter and fisher, contributing to his family's meals and provisions. His formal "Western" education ended in Grade 8. He was a canoe puller in Quw'utsun racing canoes from the age of fourteen and over the years he skippered many racing canoes, setting an example of calm, disciplined leadership that continued into the Yulhulaalh Journeys of recent years (2005 to 2017). He married and started his own family in the 1960s. To support his family he became a logger, learning much about the trees and forests from his keen powers of observation. In the 1970s, he started his employment with the Cowichan Band (now Cowichan Tribes), working on various land- and culture-related contracts. As his family grew, he realized increasingly how important his knowledge of language, culture and environments was and would be to future generations. He has dedicated the last few decades to ensuring the survival of the Hul'q'umi'num' language and documenting, through any means possible, his traditional knowledge of plants and environments so that it continues into the future.

On June 5, 2007, Luschiim received an honorary Doctor of Letters degree at Malaspina University-College (now Vancouver Island University) in recognition of his tremendous contributions to the teaching of Coast Salish culture and traditions, as well as his commitment to environmental sustainability and to preserving the Hul'q'umi'num' language. In her nomination of Luschiim for this honor, anthropologist Helene Demers stated, "He has dedicated his life to preserving and protecting his culture, language and the environment for generations to follow and he shares his knowledge generously with others. He truly serves as a bridge between First Nations and non–First Nations people. I can think of no one who more embodies the spirit of respect for diversity and lifelong learning so valued at Malaspina University-College."

Luschiim is a special and unique man; not only does he hold exceptional knowledge about the plants, language, culture and environments of the Quw'utsun people, but he is a kind, generous and distinguished teacher and Elder, who has learned what he knows from primary knowledge holders of past generations, and through extended time spent on the ocean, rivers and lakes, and in the forests, prairies and woodlands of his home place. He is committed to passing on this knowledge in a good way, for the benefit of all of us, but especially of his people. With his teachings and his vision, he is an inspiration to so many. He holds to the message passed on to him by his Elders: "Learning never comes to an end. Keep expanding your knowledge in all areas. Do all the things you want to do when you are able and comfortable, but don't ignore or neglect your families in the process" (Vancouver Island University, 2007).

NOTE: Luschiim also holds powerful and important sacred and spiritual knowledge regarding certain plants and ceremonies of his people. This knowledge about plants in sacred contexts has not been recorded in this project; it will be passed down only to those with the rights and training to be able to receive it and use it safely and appropriately.

Nancy J. Turner, January 2021

LUSCHIIM'S PLANTS

THE PLANTS, ALGAE AND FUNGI IN THE following sections are listed within their broad botanical groupings. These are plants that Luschiim is familiar with, has been told about by his own Elders, and has had experience using over the course of his lifetime. As noted in the introduction, the information has been compiled from many interviews and field outings taking place since 2005.

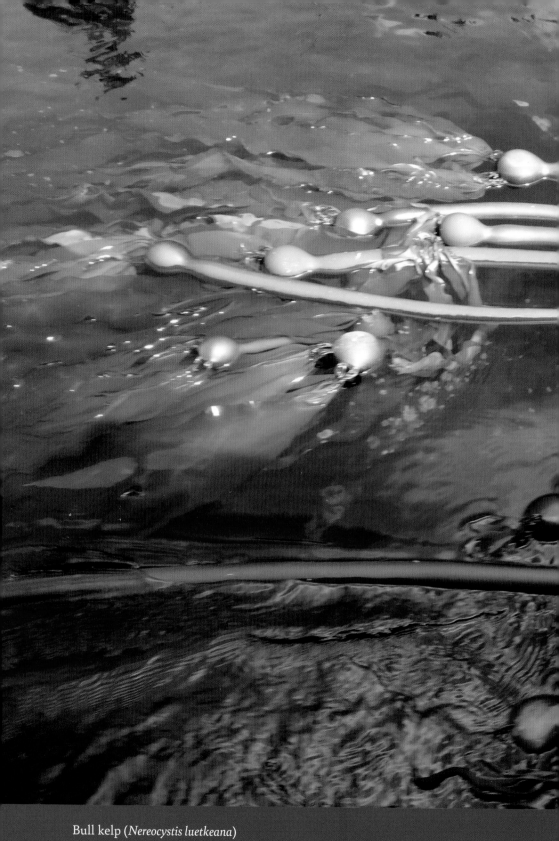

Bull kelp (*Nereocystis luetkeana*)

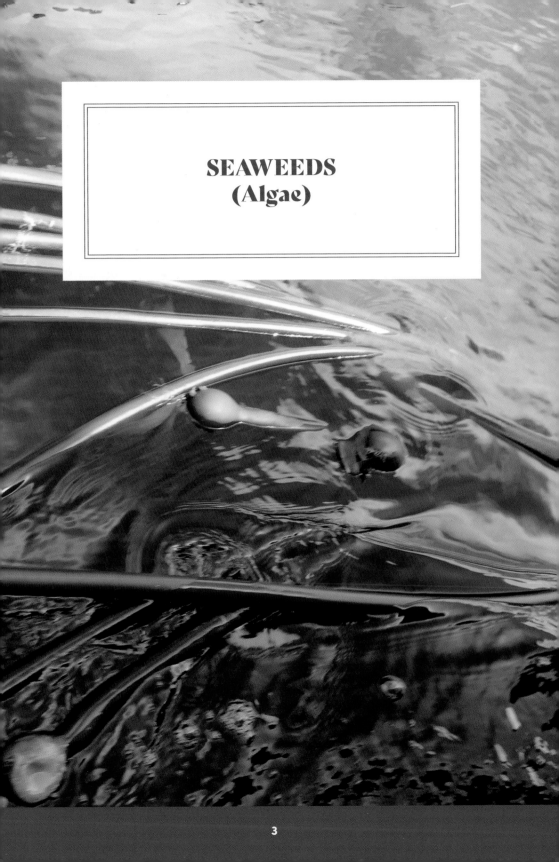

SEAWEEDS
(Algae)

SEA WRACK, BLADDERWRACK or ROCKWEED (*Fucus distichus* and related species)

FUCACEAE (ROCKWEED FAMILY)

HUL'Q'UMI'NUM' NAME: *Qw'aqwuqw*

DESCRIPTION: A short, tough seaweed in the Brown Algae group, attached to rocks with a small holdfast. The stipes are flattened with a distinctive midrib, ending in twin swollen receptacles that, when ripe, are filled with a clear gel.

WHERE TO FIND: Common on rocky shorelines in the intertidal zone from Alaska to California.

CULTURAL KNOWLEDGE: The swollen receptacles pop when you step on them. (See *tl'emukw'um*: "pop.") Luschiim explained: "Same with the **qw'aqwuqw**, sea wrack. That's the ones that pop. So you look for the younger ones to eat. But it's also used to bathe the canoe, or yourself ... Use all of it ... Squish it up. It foams up like soap" (December 7, 2010). Some people rub the gel from the broken receptacles on cuts, infections and burns.

Rockweed or bladderwrack (*Fucus* sp.) showing swollen receptacles

BULL KELP (*Nereocystis luetkeana*)

LESSONIACEAE (KELP FAMILY)

HUL'Q'UMI'NUM' NAME: *Q'am'*

DESCRIPTION: A tough kelp (Brown Algae) with a long, cylindrical stipe attached to the ocean floor by a stout holdfast, hollow at the upper end and culminating in a large, hollow "bulb" bearing 50 or more flat, elongated blades, or fronds.

WHERE TO FIND: This distinctive tubular seaweed grows from rocks in the subtidal zone, forming dense patches, or "kelp forests," often 30 metres or higher, which are important habitat for many marine species. It is common all along the Pacific Coast from Alaska to California.

Bull kelp (*Nereocystis luetkeana*)

CULTURAL KNOWLEDGE: Luschiim described this long-stemmed kelp:

> Long stems going down to the bottom, it's got lots of hair. That's *q'am'*.
> I heard a name [for the large bulb part] but I was very small, and I
> don't remember it. And I heard the name for ... the leaves; we would
> just call it *she'yutun*, "hair"... The *she'yutun* of the *q'am'*, we used it to
> keep your [fish] catch cool. You laid some of that on the bottom of your
> canoe. The canoe had boards to keep your catch off of the canoe, off of
> any water that comes in, put a layer of that *she'yutun* or other seaweed
> such as eelgrass. You lay your catch in there, then ... depending on the
> amount of sunshine, you put a big layer of that *she'yutun* of the *q'am'* on
> top. To keep your catch cool. (December 7, 2010)

The stalk of the kelp is called *shtl'up'isnuts* ("tail"). Luschiim explained its use: "The bigger part of the *shtl'up'isnuts*, which is hollow, the 'tail' part, was used for a toy, in a game—for whipping, or playing with each other, snapping it. It's also used for other things. I never did use it, but I was told it was used to braid into a cord, a long rope" (December 7, 2010).

The larger hollow stipes had other uses:

The hollow, bigger ones you could find, we used for curing things such as a bow, *taxwa'ts*, bow and arrow. You whittled your bow [usually of Pacific yew wood] into the shape you want ... and then you put your bow into that hollow [stipe], along with some other herbs ... then you had a fire going just like you would when you're going to do clams; the ground is nice and hot. You close up both ends, you put your bow inside that hollow, you stick it in there, you leave it there for a couple of days until the ground is cooled off and you just treat it with other herbs, probably some oil. They utilize it there. That's to keep the bow flexible and limber, springy. So, that was used for that, the big part of the tail of the *q'am'* ... so whenever you can find big ones that are something like 3.5 to 4 inches [across], you slip your bow in there ...

So, when you're harvesting out in the sea, you go by the bull kelp, big. You got your long spear here, your *shtl'e'lhunum'*, your three-pronged spear ... long one, 18 feet, some long, some short. You ... grab a big pile of that, the head and the hair of the kelp ... and you wrap some of that across your canoe; you're not going to go anywhere. And you look down. You're going to smash up a bunch of those sea urchins ... smash them up and keep the point of your spear close to it. And then you look for the rockfishes, maybe a rock cod will come to eat there, ling cod, and you spear them ... and later on, when the gas boats came, those kelp beds were big enough that you could also tie your [gas boat]—I guess there were smaller gas boats—like 25 to 30 footers would be a big one in the early days—you could tie up on there too with your gas boat.

Ah, when the water's quite rough, you can, on some places on the coast, you can go inside the kelp beds, meaning between the shore and the kelp beds, to get away from some of the rough water.

> Or if the current's against you, get away from the swift water, so ... (December 7, 2010)

Luschiim stated that people have probably been anchoring their canoes and boats to kelp for thousands of years. He also recalled how kelp was used to pull their canoe up onto the beach:

> "As kids—and my dad helped with this—it was our job to pull a canoe up onto the beach, either way up or put it back in the water. If the canoe wasn't that heavy and we were able to use just some of the [kelp] washed up on the shore. We seen kelp rounds, the bigger parts of the rounds, we lay it like poles to drag the canoe up and down." (December 7, 2010)

Luschiim also mentioned that the kelp beds were good indicators of other species, such as sea urchins and a fish with bluish spots that was harvested for rendering its fat: ratfish. He also noted that you can eat part of *q'am'*:

> You can eat the tender parts of it if you have to. You can be blown out ... you're out there, you get out there somewhere in the Gulf Islands, or wherever out there, and the seas come up, big winds, and you have to be stuck out there for days and days. You have to survive somewhere. Maybe you're not ready to be stuck for three or four days, [so] you can eat that. (December 7, 2010)

MEMBRANOUS SEAWEEDS

(*Pyropia abbottiae*, syn. *Porphyra abbottiae*; *Pyropia perforata*; and related *Pyropia* and *Porphyra* spp.);

SEA LETTUCE (*Ulva lactuca* and other *Ulva* spp.)

BANGIACEAE (MEMBRANOUS RED SEAWEED FAMILY)
ULVACEAE (SEA LETTUCE FAMILY)

HUL'Q'UMI'NUM' NAME: ***Pulhtalus*** (see ***plhet***, "thick"): thicker, darker type, found higher up on the beach (possibly *P. perforata*, and/or *P. abbottiae*); and ***lhuq'us***: thinner, green type growing farther down on the beach

DESCRIPTION: Luschiim recognizes two different types of "seaweed," possibly different species of *Pyropia*, or different genera—*Pyropia* and *Ulva*. Species in

Red laver seaweed (*Pyropia abbottiae*)

both of these genera are membranous, like thin sheets of rubber. *Pyropia* species are classified in the Red Algae group, and *Ulva*, or "sea lettuce," which is bright green in colour, is in the broad group of Green Algae.

WHERE TO FIND: These seaweeds are found attached to rocks in the intertidal zone; there are various genera and species that are difficult to tell apart. Species in these genera occur all along the northwestern Pacific coast, from the Aleutians to Baja California.

CULTURAL KNOWLEDGE: According to Luschiim, **lhuq'us** refers to a greenish seaweed (called "number ones"). Luschiim described the difference between the two types: "The dark one is thicker. Some call it red, some call it brownish ... It's higher up; it grows at a different part of the beach than the

Sea lettuce (*Ulva* sp.)

green one. And we call it **pulhtalus** ... It's thicker than the green one (note that **plhet** means "thick," and relates to this name)" (December 7, 2010).

Luschiim noted that the opposite of **plhet** is **ts'umiil'** ("thin"). The name **pulhtalus** may refer to *Pyropia perforata*, purple laver, which grows in the upper to low intertidal zone. Luschiim said that the Quw'utsun did eat both **pulhtalus**, the darker, thicker one, and **lhuq'us**, the green one. He said that he personally hadn't eaten them, but those of a previous generation had eaten both types:

> Possibly the last one who really done lots [picked and dried the seaweeds] was my dad, who used to really do that lots; my dad picked them, which was in the 1920s. My uncle Ernie picked them, and he was born in 1916. My dad's cousin picked them, and he was born in 1919. Those that lived on the islands—Kuper, Chemainus—they went out and picked them. We're kind of away from the sea here, so we went farming while the others stayed with living off the sea. (December 7, 2010)

Lung lichen (*Lobaria pulmonaria*)

LICHENS

Various species of tree lichens, especially LUNG LICHEN (*Lobaria pulmonaria*); OLD MAN'S BEARD LICHENS (*Usnea* spp., *Alectoria* spp.); SILVERY TREE LICHENS (e.g., *Cetraria* spp., *Evernia prunastri*, *Hypogymnia* spp.)

LOBARIACEAE (LUNG LICHEN FAMILY)
PARMELIACEAE (PARMELIA FAMILY: OLD MAN'S BEARD AND TREE LICHENS)

HUL'Q'UMI'NUM' NAME: ***Smuxt'ulus*** (lung lichen), ***she'yutun*** (literally "hair"; used for old man's beard lichens, horsehair) and ***squq-p'iws tu p'hwulhp*** (literally "stuck on oak tree"; used for oak lichens), or ***q'uts'i'*** (general name for mosses and hair-like lichens)

DESCRIPTION: Lichens are composite organisms, with two or more algae and fungal species growing so closely together that they have a unique shape, colour and ecological characteristics.

WHERE TO FIND: Many different kinds of lichens occur in southeastern Vancouver Island, growing on tree branches and trunks, on rock faces or on the ground.

CULTURAL KNOWLEDGE: Luschiim described several different types of lichens, including lung lichen (*Lobaria pulmonaria*):

> The ones on the maple, they're green, and the silvery grey on the underside [lung lichen]. That's ***smuxt'ulus*** ... And so, the one that grows on maple is ... the preferred one of that kind ... You use it to make tea, or we add it to other tea ingredients. It's also a medicine. We mainly use it for tea. We were told it's an energizing drink ... Today, the equivalent would be Gatorade ... power drink ... We mixed ours with ***ququn'alhp***, which is your prince's pine [pipsissewa, *Chimaphila umbellata*].

So that's **smuxt'ulus** ... Get 'em off a live tree. If you are going to get them after a big wind, you go the next day, not any later. If you go the next day there'll be lots. You make sure they're up off the ground, because once it hits the ground it gets attacked by bugs ... [so get the ones fallen on] bushes, or maybe a big branch [is] broken, it gets hung up. (December 7, 2010)

Luschiim talked about another type of lichen with the same Hul'q'umi'num' name, which grows on ironwood (*Holodiscus discolor*):

There's another one, the one that grows on ironwood, **q'eythulhp** ... [The] **smuxt'ulus** that grows on **q'eythulhp** ... is a medicine. I wasn't told what kind it is, but it's a medicine, very special. They don't grow that way everywhere. One of our favourite places was on the north shore of Shawnigan Lake: that mountain there, the one with hardly any trees on

Lung lichen (*Lobaria pulmonaria*)

the top, between the top and the lake. They go to the steep part [and] all the way down is that **q'eythulhp**, the ironwood, around there. They know that. So there's certain places are known to have certain things, that's one of them. (December 7, 2010)

Then Luschiim discussed the light green, hair-like "old man's beard" lichens (*Alectoria* spp.; *Usnea* spp.; also sometimes called by the general name for moss, **quts'i'**), like the ones growing on the cottonwoods outside of the window of his home, which are named after the trees they are growing on:

[You] just say **she'yutun** ["hair"] of that particular tree: "hair" and then name that particular tree. Some of it is medicine, depending on [the type]. If it's really cold, it's gotta be nice and dry, good for fire. A certain kind—I don't know which one—is good for a band-aid ... I was told they have healing properties, so to me, that would be like an antiseptic ... (December 7, 2010)

Old man's beard lichen (*Alectoria sarmentosa*)

Another type of lichen Luschiim described is called ***squq-p'iws tu p'hwulhp***, one that grows on Garry oak trees (*Quercus garryana*). He explained the reason for the name, and the way this lichen is used:

> If this [object] is stuck on me, that's ***squq-p'iws***, so, ***squq-p'iws*** is something that's stuck on [to something]; that's like the lichen that's stuck on the Garry oak [***p'hwulhp***]. ***Squq-p'iws tu p'hwulhp*** ["what is stuck on the Garry oak"] ... it's used for ***tth'iq'ula'th***. I think it might be called "thrush," on a baby; problems with their mouth ... The doctors take care of that now at the hospital, so we don't get to use it anymore, but that's what it's good for ... newborns. (December 7, 2010)

This lichen may be one or more species of *Hypogymnia* spp. or *Cetraria* spp. or *Evernia* spp.—all lichens that grow on the bark of trees. Luschiim also noted that "***squq-p'iws*** is a generic word, could be [growing] on a rock, or could be on other trees. So there is others that I don't know about, growing on other trees or rocks; you have what it's growing on in the name named in that phrase ..." (December 7, 2010). In January 2010, he pointed to a species of *Parmelia* and said, "Flat, wrinkled up leaves, leafy. ***Squq-p'iws***. So here's a good example of ***squqep'*** ... stuck on. So ***squq-p'iws*** is just stuck on lightly on the bark ... Again, that's for the babies with ***stth'iq'ula'th*** [thrush], so it's an antiseptic" (January 23, 2011, and November 2017).

Luschiim collects lichens for medicinal use at the Cowichan Garry Oak Preserve.

Red-belted bracket fungus (*Fomitopsis pinicola*)

FUNGI
(Including
Mushrooms)

TREE FUNGI, BRACKET FUNGI or SHELF FUNGI (various species, including *Ganoderma applanatum*, *Fomitopsis pinicola* and *Polyporus* spp.)

VARIOUS FAMILIES, INCLUDING GANODERMATACEAE, FOMITOPSIDACEAE, POLYPORACEAE

HUL'Q'UMI'NUM' NAME: *Hwuhwa'us-uhw* (layered tree fungus) or *tuw'tuw'eluqup* (literally "echo," and also used for "telephone"; Luschiim, April 16, 2015)

DESCRIPTION: There are many different types of tree fungi, mostly rounded with brownish, buff or dark tops and white undersides. Most are woody and tough and vary in texture.

WHERE TO FIND: These fungi usually grow on rotten logs, snags or sometimes on the trunks of still-living trees, both evergreen and deciduous, depending on the species of fungus. They are widespread in our forests.

CULTURAL KNOWLEDGE: Luschiim noted that **hwuhwa'us-uhw** is a type of **tuw'tuw'eluqup**, which is a more general name for tree fungi (April 16, 2015). The ones called **hwuhwa'us-uhw** grow in layers on the tree trunk. He talked about the association between the tree fungi and echoes, and also about their use as a medicine:

> ... **tuw'tuw'eluqup** is in charge of the echoes. That's its job here on the earth, to look after the echoes. Some of them have very good, um, medicinal properties. Again, **tuw'tuw'eluqup** growing on the **p'hwulhp** [Garry oak], on the still growing green trees, is said to be very good for cancer as a drink, or if need be, you can just gnaw on it ... (December 7, 2010)

He also talked about different kinds of tree fungi:

Um, some of them ***tuw'tuw'eluqup*** that grow on certain trees is used for red paint. You bake it. Nowadays you can wrap it in tin foil and bake it in the oven, yeah. And it turns red ... One of them is off the big willow [Pacific willow, Sitka willow], the ones that grows into a big shrub or a small tree, meaning the base could be up to two feet or so and the height would be something like, maybe 30 feet. The other one grows on white fir [probably amabilis fir]. (December 7, 2010)

One fungus that yields red paint is "Indian paint fungus," *Echinodontium tinctorium*; this is probably the species Luschiim refers to as growing on a "white fir." Luschiim said you grind it to a powder after it is burned, then mix it with a medium to use as a paint (April 18, 2016).

Bracket fungus, or artist's fungus (*Ganoderma applanatum*)

MUSHROOMS (many species)

HUL'Q'UMI'NUM' NAME): *kwumsuli'qw* (general name for mushrooms, referring to "hat"; see *–iqw*, "head;" see also the Central Salish root **q'əməs**, "mushroom, fungus"—Kuipers, 2002)

DESCRIPTION: There are hundreds of different kinds of mushrooms in British Columbia, of many shapes, sizes and colours. The Hul'q'umi'num' name is a general one that covers all of them and refers to the "cap" (conical, rounded or flat) that is usually on top of a cylindrical stem, or stipe. Under the cap are thin radiating gills, or sometimes pores, or small holes where the spores are produced. The stipe may have a ring around it, and sometimes a cup at the base, especially in some of the poisonous types, the *Amanitas*.

WHERE TO FIND: Mushrooms grow in a wide variety of habitats, from open fields to forested areas, usually in the fall or spring, following a rainy spell.

Questionable Stropharia mushroom (*Stropharia ambigua*)

Some mushrooms grow on sawdust or wood, others on soil. They are the spore-bearing part of a fungus that has a widely spreading, fine root-like mycelium in the soil or other substrate. Usually the mushrooms themselves are soft and short-lived.

CULTURAL KNOWLEDGE: Luschiim noted that ***kwumsuli'qw*** is a general term for many different kinds of mushrooms and mentioned one type that grows in cow pastures, the "crazy mushroom" (magic mushroom, *Psilocybe*), which is hallucinogenic. He knows that some people pick wild mushrooms to eat, but he does not, "because it's easy to pick the wrong one without knowing it." He said very few people ate mushrooms in the old days (December 7, 2010).

Waxy laccaria (*Laccaria laccata*)

ORANGE JELLY FUNGUS, or WITCH'S BUTTER

(*Tremella mesenterica* and related species)
TREMELLACEAE (*TREMELLA* FAMILY)

HUL'Q'UMI'NUM' NAME: *Shmut'qw'*

DESCRIPTION: A shiny, bright yellow-orange, lobed and convoluted gelatinous fungus, somewhat brain-like in appearance. It is similar to another species, *Pseudohydnum gelatinosum*, also called "witch's butter," but more yellowish. Both are edible but tasteless.

WHERE TO FIND: Orange jelly fungus typically appears after a heavy rain, from May through November. It occurs throughout North America. It lives off dead or decaying plant material and is usually seen on rotting wood.

CULTURAL KNOWLEDGE: Luschiim described how people used the orange jelly fungus as a thirst quencher if they were out with no water available:

> We did [eat bitter cherries] ... they're quite bitter, but if you're up in the mountains and there's no water ... (In my early days, water containers were a rare thing. The only water containers were whiskey bottles and like that.) When there's no water [in the summer], you learn how you can get moisture up there. That's one of them [that they used]. The other two are your Oregon-grapes [and] ... ***shmut'qw'***—I think it's called the orange jelly fungus ... there's three different colours [of the jelly fungus]. The first one, the orange, comes in different shades of orange, [then] you have your yellow [possibly *Tremella mesenterica*], and then you have your white [*Pseudohydnum gelatinosum*], so there's three colours. The ones we used were the ones growing on a rotten stump, rotten wood. So ***shmut'qw'*** I imagine you have to sort of acquire the taste. And for purposes of being very thirsty and trying to quench your thirst, it was a welcome thing. So we were used to it, but I know others who tried it and they couldn't [eat it] ... ***shmut'qw'***. I got that name from Edward Jack. You've heard of Felix Jack from Mayne Island? That's his grandson ... my cousin. Felix Jack, his name will come up another time, talking about mint. ***Shmut'qw'***, jelly fungus. (January 23, 2011)

Green peat moss (*Sphagnum girgensohnii*)

MOSSES AND LIVERWORTS
(Bryophytes)

MOSSES (various species)

HUL'Q'UMI'NUM' NAME: *Q'uts'i'* (general name for mosses and mossy lichens)

DESCRIPTION: Mosses and liverworts are small, bright green plants without true flowers or leaves, which often grow in dense patches or clumps.

WHERE TO FIND: These spore-bearing plants usually grow in damp, shady locations, on the ground, on rocks or tree trunks, or in wetlands such as bogs. There are hundreds of different kinds, only of few of which have specific cultural uses.

CULTURAL KNOWLEDGE: Luschiim calls mosses in general *q'uts'i'*, whether they are growing on the ground or on tree trunks or rocks. He noted that some types grow in the water as well. He said that some of the different types

Common mosses (*Dicranum scoparium* and *Rhizomnium glabrescens*)

used to have special names. He talked about a special type that grew in the water, which was used for some kind of medicine:

> Some of them are very good medicine. One that grows in the creeks, in the water, very good medicine, but I don't know what for. But my mom had to harvest it back in the early 1930s ... She got them in the summertime, in berry-picking time, at a place called Camp 10, I believe on the north shore [of Cowichan Lake, beyond Youbou] ... (December 7, 2010)

Sphagnum mosses, growing in peat bogs and other moist, acidic soils, are particularly absorbent and were formerly probably used for baby diapers.

Common green sphagnum moss (*Sphagnum girgensohnii*)

Sword fern (*Polystichum munitum*)

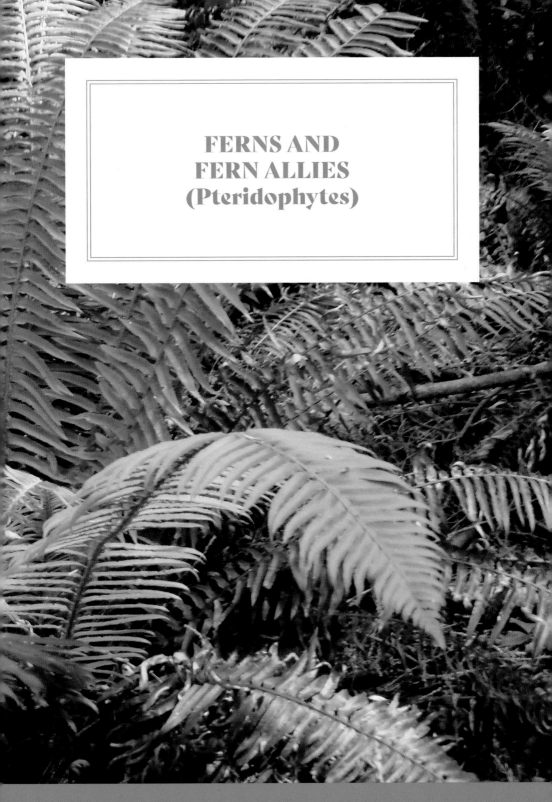

FERNS AND
FERN ALLIES
(Pteridophytes)

SCOURING RUSH, or BRANCHLESS HORSETAIL (*Equisetum hyemale*)

EQUISETACEAE (HORSETAIL FAMILY)

HUL'Q'UMI'NUM' NAME: *Hwkw'ul'u*, *hwkw'ul'a* (see *kw'ulu*, "stomach")

DESCRIPTION: These plants have dark-green segmented stalks and a "scratchy" texture due to the presence of silicon in their cells. Usually the segments are hollow and sometimes are filled with fluid. The spores are borne in a small, dark cone-like structure at the tip of the stem.

WHERE TO FIND: Scouring rushes grow in patches in moist, open places such as the edges of lakes and ponds, often in sandy or gravelly soil. They are widespread.

CULTURAL KNOWLEDGE: Scouring rushes were used like sandpaper and also to stimulate and enhance the voice:

Good for polishing, for your knitting needles or maybe an arrow. You get that and you rub your stem or needle or whatever it might be. Very fine, shiny. That's one use. Then they use the black end? You chew that, swallow the juice and spit out the pulp. Only the solid ones. There's lots of black tips. Many of them are ... hollow. There's only a few that are solid. You look for the solid ones. For singing ... for singing songs. You might stand there and sing all night. It helps their voice box or whatever. (Luschiim, December 7, 2010, and October 2017)

GIANT HORSETAIL (*Equisetum telmateia*)
and COMMON HORSETAIL (*E. arvense*)
EQUISETACEAE (HORSETAIL FAMILY)

HUL'Q'UMI'NUM' NAME: *Sxum'xum'*

DESCRIPTION: These horsetail species are both herbaceous perennials with separate vegetative (green) and spore-bearing (pale whitish) stalks; common horsetail (*Equisetum arvense*) is usually about 30 to 50 centimetres tall, while giant horsetail (*E. telmateia*) can reach 1.5 metres. The stems are hollow and jointed, the leaves reduced to rings of bracts at the stem nodes or joints. In both species, fertile stems, producing spores in elongated cone-like structures, appear in spring before the green vegetative stems, then die back. The green, sterile stalks, with regular whorled branches, expand and continue to live for the rest of the season, then die back in the fall.

Giant horsetail (*Equisetum telmateia*) spore-bearing shoots

WHERE TO FIND: These horsetails grow in damp soil, around springs, swamps and seepage areas, and in open woods, often forming large patches from underground rhizomes. They are both very common on Vancouver Island.

CULTURAL KNOWLEDGE: The young shoots of the giant horsetail are broken off and eaten as a springtime food:

> We only know them as shoots and that's just **sxum'xum'**, I've never heard a name for the fully grown [ones] ... there's different kinds. The only ones we eat were the ones that are kind of golden-coloured ... the sort of golden ones, maybe shorter than four inches, just out of the ground. Yep ... They get tall. You just eat them raw ... They're pretty watery. The top is pretty watery, pulpy, so we didn't eat it ... (Luschiim, December 7, 2010)

The mature, green plants of **sxum'xum'** were not eaten, and are actually poisonous. The spore-bearing shoots of common horsetail (*E. arvense*), which are generally much smaller and more slender, may have also been eaten, but not as commonly as those of giant horsetail.

Giant horsetail mature vegetative stalks

GOLDENBACK FERN

(Pentagramma triangularis)

PTERIDACEAE (MAIDENHAIR FERN FAMILY)

HUL'Q'UMI'NUM' NAME: Not recalled by Luschiim

DESCRIPTION: A small, somewhat wiry fern with clustered fronds, the leafy part dark green with a general triangular shape and the stems black and wiry.

WHERE TO FIND: Grows on rocky outcroppings and in crevices, such as on Koksilah Ridge and Mount Tzouhalem.

CULTURAL KNOWLEDGE: Luschiim recognized this fern as one that was used to make a colourful design on baskets. He said it was similar in some ways to maidenhair fern, but while maidenhair fern grows right in the spray beside waterfalls, this one grows on drier rocky bluffs (April 16, 2015).

LICORICE FERN

(*Polypodium glycyrrhiza*)

POLYPODIACEAE (POLYPODY FAMILY)

HUL'Q'UMI'NUM' NAME: *Tl'usiip*; licorice fern on a broadleaf maple: *tl'usiip-s tu ts'alhulhp*; licorice fern on a red alder: *tl'usiip-s tu kwulala'ulhp*; licorice fern on a rock: *tl'usiip-s tu smeent*

DESCRIPTION: A small fern, with fronds pinnately once-divided, growing in patches from branching, greenish, licorice-flavoured rhizomes; fronds are

evergreen in winter, dying back in early summer and regenerating by late summer. The spores are borne from bright orange spots on the undersides of the fronds.

WHERE TO FIND: This fern grows in dense patches on mossy tree trunks and mossy rock faces. Common throughout coastal British Columbia.

CULTURAL KNOWLEDGE: The "roots" (rhizomes) of this fern contain a very sweet-tasting compound and are used variously for medicine (Luschiim, December 7, 2010, and April 16, 2015). Luschiim noted, "It has different applications depending on where it's growing." For example, you would say ***tl'usiip-s tu ts'alhulhp*** for licorice fern growing on a maple tree, ***tl'usiip-s tu kwalala'ulhp*** if it is growing on a red alder and ***tl'usiip ni' 'utu smeent*** ("it grows on the rock") if it is growing on a rock face.

He explained how it was used:

> So the ones we use for ... the voice is the ones that grow on maple or alder. We get the roots; they can travel on the bark. Depends on what you use it for. You chew on that and swallow the juice ... And the one that grows on rock, it's the same: ***tl'usiip ni' 'utu smeent*** ... You get those root parts, use it to make tea. Could be a general tea or it could be something that kind of curbs the appetite. It's supposed to work ...

Notably, ***tl'usiip*** rhizomes are also widely used to treat colds, coughs and sore throats, and as a sweetener for other kinds of medicine (for example from tree barks) that are bitter tasting. They are simply chewed, and the juice swallowed, or are boiled in water to make a concentrated tea.

SWORD FERN *(Polystichum munitum)*

POLYPODIACEAE (POLYPODY FAMILY)

HUL'Q'UMI'NUM' NAME: *Sthxelum* (Luschiim, December 7, 2010, and October 2017; see also Proto–Coast Salish *s-(ts)xal-m*, "sword fern"—Kuipers, 2002)

DESCRIPTION: This fern is evergreen, with long, dark-green fronds growing together in a large clump from a thick rootstock. New fronds emerge and mature before the previous year's fronds start to die back. The frond is once-divided, with numerous pointed, toothed leaflets growing along the stalk in a feather-like arrangement. As in all ferns, the young fronds are curled as "fiddleheads." Spores are produced in small, brownish structures arranged in lines on the undersides of some fronds.

WHERE TO FIND: Sword fern is predominantly a coastal species and is found throughout Quw'utsun territory, thriving in rich moist soil under conifers or alders and maples.

CULTURAL KNOWLEDGE: This is a very sacred plant, used ceremonially by Coast Salish peoples. Luschiim described it: "Sword fern, *sthxelum* ... The one that grows in clumps ... It's used all year round, including the winter dance" (April 16, 2015). Earlier, he noted:

> Probably in January, depending on the elevation, maybe even December, you go to that big clump, you dig in there and you get the fiddleheads. That's good. But once you can see them in that clump, it's too late. Yeah, you just dig into it. You eat that, yes. Nice and juicy. If you're a little bit too late, it gets kind of hairy. Yes, also used by the *st'alkwlh* non-dancers. It's used for *thul'shutun*, something like a foot mat, a mat for your feet, when you go to *kw'aythut* (bathe). You get that and lay it [down]; that's what you're going to step on, on the ground. When you finished, you're supposed to take those ferns [and] put it up where the wind can catch it, in the branches ... so the wind will cleanse off whatever you've wiped off around there. (June 15, 2007, December 7, 2010, and October 2017)

Luschiim noted that the shoots could be used as food. He had heard of sword fern fronds being used in pit-cooking but has never used them himself; he would use bigleaf maple leaves, bracken fern and fireweed.

BRACKEN FERN (*Pteridium aquilinum*)

DENNSTAEDTIACEAE (HAY-SCENTED FERN FAMILY)

HUL'Q'UMI'NUM' NAME: Fronds: ***suqeen*** (possibly related to ***seq***, "outside," and ***-e.n***, "ear"); "roots" (large black rhizomes): ***se'uq*** (Luschiim, December 7, 2010; see also Proto–Coast Salish ***saʔq***, "bracken fern (root)"—Kuipers, 2002); also possibly ***ptákwum*** (see Proto–Coast Salish ***ptakʷ***, "bracken fern"—Kuipers, 2002)

DESCRIPTION: This is a large fern, growing to one metre or taller, growing in patches, with long-stemmed fronds arising singly from a deep, branching rhizome. The leafy part of the frond is broadly triangular, divided into small segments, giving the fronds a lacy appearance.

WHERE TO FIND: Bracken is common, often growing in dense patches in prairies, clearings, road edges and open woods, in well-drained soils.

CULTURAL KNOWLEDGE: Luschiim was told that the Nooksack people are named after this fern: ***hwse'uq***. He commented, "They used to have lots and lots of bracken fern in their territory; people went there to harvest it.

Bracken fern rhizomes

They believe that if you don't pulverize it really well [***teehwt***, "to pulverize"], it will turn into snakes ..." (September 24, 2010).

"Yes, there used to be *lots* over there ... They're named after the bracken roots ... And that information comes from the people at Nooksack [Mount Baker Road]. They may say it a little bit differently from me" (December 7, 2010, and April 16, 2015).

Bracken fern has many important qualities:

Bracken fern fronds

> The root of the thing is called the **se'uq** ... **Suqeen** is used for many things, like a thing to lay on if there's a whole bunch of it ... We used it to clean out the blood of a deer or elk. Also for getting fish ... if you're doing lots of fish, like hundreds of fish ... you got lots of slime and blood. It is the kids' job to go and collect it, maybe keep changing it ... get the slime away ... You go to start your fire for smoking a fish, when you first get your fire going ... there's a pile of this used **suqeen**; after you get the fire going you use this as part of ... the flavouring ... The **se'uq**, the root is ... what we used to use to make a flour ... out of it. (Luschiim, April 16, 2015)

Luschiim also described using bracken fern, maple leaves or fireweed to stuff the cavity of a freshly killed deer when it is being cleaned. This process seasons the meat and gives it a good flavour (see page 90).

Grand fir (*Abies grandis*)

CONIFEROUS TREES

GRAND FIR (*Abies grandis*); AMABILIS FIR, or SILVER FIR (*A. amabilis*); and SUBALPINE FIR (*A. lasiocarpa*)

PINACEAE (PINE FAMILY)

HUL'Q'UMI'NUM' NAME: Grand fir is called **t'a'hw**; it is the most common of the "true fir" species in the Cowichan area. Amabilis fir, Luschiim said, can be called **stth'ux t'a'hw** ("silver grand fir"; see **stth'ux**, "silver"); subalpine fir is called **tsa'luqw t'a'hw** (literally "high elevation grand fir"; see **tsa'luqw**, "high elevation"; December 7, 2010); pitch (of all three species), produced in blisters on the bark of young trees: **smuqw'iws**; grand fir blisters: **smuqw'iws tu t'a'hw**.

DESCRIPTION: Grand fir is a tall, straight tree growing to 80 metres high. In young trees the bark is smooth and greyish, often dotted with resin blisters,

Grand fir (*Abies grandis*)

which contain strong-smelling liquid pitch. It becomes rougher and more ridged with age. The needles are flat and blunt, notched at the tip, dark green above and with two white lines beneath. The branches are flat with spreading needles. The large, upright, brown seed cones are borne near the top of the tree and release seeds by breaking apart rather than falling as whole cones. Amabilis fir and subalpine fir are similar, but amabilis fir has a dense row of short, forward-pointing needles along the tops of the twigs. Subalpine fir needles are white on both sides and tend to spread around the top half of the twigs. The cones of amabilis fir and subalpine fir are purple.

Grand fir bark with raised pitch blisters, characteristic of this and other *Abies* species

WHERE TO FIND: Grand fir grows, often with Douglas-fir, on the drier side of Vancouver Island and on the Gulf Islands, throughout Quw'utsun territory. Amabilis fir and subalpine fir grow at higher elevations, and amabilis fir grows on the wet west coast of Vancouver Island.

CULTURAL KNOWLEDGE: Luschiim explained that when the salmon is still silver, coming up the river, it is called *stth'ux*: "silver." So he applied this name to amabilis fir, which is sometimes called "silver fir." He noted that amabilis fir and subalpine fir are "a little bit more potent [than grand fir] in the smell and the taste," but that the uses of all of these are similar. He noted that the original names for amabilis fir and subalpine fir have been forgotten, so these are "new words," descriptive of these trees (December 7, 2010).

Luschiim talked about grand fir, *t'a'hw*, in more detail:

The grand fir ... has quite a few uses. One of them is for a dye. To colour material a darker colour ... not really black, but it's dark. So you get the bark and the branches and boil it up with whatever you're going to make darker. [It is] also used as one of the ingredients for making steel, such as traps, kind of rust proof. So you get the bark of *t'a'hw*, the bark of *ts'qe'ulhp* [white pine], *kwalala'ulhp* [red alder bark] and possibly the cones of the red alder, little round cones, or when they're brown [ripe], a little more open. And you put them all together and brew them up and put your traps in there ... Two things. It rust-proofs it somewhat, but it also takes away the human scent. So that's in addition to making material darker, you can use it for that purpose also [as a coating so it doesn't rust].

For the *t'a'hw* branches, you can use that for hair shampoo ... [and] hair tonic. For one person you can get the ends of four branches, roughly 12 to 15 inches long, for those, brew it up and it'll be enough to do your hair. [NT: As a rinse?] Yes. The *t'a'hw* tips of the branches, you can chew that anytime. It helps you to fight off cold, ordinary cold sickness, runny nose, that type of stuff. It's also spiritual, food ... the tips. For the grand fir, the *t'a'hw*, it has pitch blisters and that is called *smuqw'iws*, and that's good for ... the same thing you use cough drops for. But for an average-sized blister, small, medium, large, take a good look. The medium size, you use probably a quarter [of the pitch it contains]. If you're strong you can use half of it, but that's getting to be quite a bit. It's pretty potent. (December 7, 2010)

There are other uses for these firs:

The tips could be used—I don't know what you'd call it—it kind of clears the voice. The tips of all of them ... stronger further after they've developed, but you could use them at any time of the year. [It's] used for shampoo throughout the year. Get four branches. Over time you find out how long, how much you want, four of those, you brew it up in a pot. It makes your hair really soft. It's [also] used for deodorizer. Fumigating a house. Maybe not just deodorizer but aromatic—fragrance ... And you can bring a few branches home ... we're going hunting tomorrow, and I'll leave my clothes with the branches overnight, then I put my clothes on in the morning and away I go. To hide my own smell. That was a big thing in my early days. Boys used to do that ... in the early fifties. The teenagers of the day were still doing that. I tried going there when I was six, so that would be 1948. I was still doing that ...

Grand fir boughs could also be used as an incense, to cleanse the house, as purification ... (Luschiim, December 7, 2010)

Luschiim also talked about the little pitch blisters on the bark of these trees:

Iws, you use it to refer to something on someone's body ... ***muqw'***, if I had a blister here and I pressed it and it burst, that's ***muqw'***. ***Smuqw'iws*** is that blister of that body of this tree ... So the pitch blisters are ***smuqw'iws*** ... of all those three [grand fir, amabilis fir and subalpine fir]. Ah, they're very good to use it for, like what do you call those Halls [cough drops]? Yeah. You burst it and it'd get on your nail ... not too much. Off a medium-sized blister, you use about half. You could use more [but] ... [it is] strong! That pitch, ***smuqw'iws***, is also used in other medicines, and—I don't know all the ingredients that would go with it—you brew it up, put that pitch in there. There's many kinds of ***smuqw'iws***, so ***t'a'hw***, ***smuqw'iws tu t'a'hw***, that's the one [that] could be any one of those three.

You can harvest them; I've seen my uncle go up with a cup and a teaspoon and a stepladder. And he ... had a whole bunch in his cup ... Yeah, [to use it for skin sores], heat it up, put it on as hot as you can. It's usually mixed with some other things, but I don't know about the other things ... You can [mix

Amabilis fir, or silver fir (*Abies amabilis*) bough

Subalpine fir (*Abies lasiocarpa*) bough

it with deer fat] ... **'anuw'** of the deer ... the fat of the deer. You render it, you boil it, skim off and you end up with a pure white fat, or **'anuw'**. You can mix the two together. You can also—while we're at the **'anuw'**—you can add the sticky parts of the cottonwood in the spring, make that **'anuw'** fragrant. (December 7, 2010)

Luschiim said that in some places, the grand fir needles are particularly strong and aromatic, and some people with asthma can use them as a substitute for a "puffer"; they help to clear the lungs and bronchial passages (May 25, 2017). Luschiim noted that people learned to use grand fir (and arbutus leaves) for medicine for wounds by observing a deer (a big buck) with gunshot wounds. The hunter tracked it over the snow and observed that its wound was treated by another deer that came to its aid and applied this medicine. (September 24, 2010)

YELLOW-CEDAR (*Callitropsis nootkatensis*, syn. *Chamaecyparis nootkatensis*)

CUPRESSACEAE (CYPRESS FAMILY)

HUL'Q'UMI'NUM' NAME: *Pashuluqw*; inner bark: *sluwi's tu pashuluqw* (Luschiim, December 16, 2010)

DESCRIPTION: A cypress-like tree growing up to 50 metres tall, with a somewhat drooping appearance. The trunk often twists slightly, and older trees have a ragged appearance. The greyish bark has long vertical furrows, similar to those of red-cedar. The boughs are bluish green. The pollen cones are very small, and the seed cones are globular and about the size of a large pea, releasing small winged seeds when they mature.

WHERE TO FIND: Yellow-cedar is usually found in moist areas at upper elevations, including avalanche shoots, ridge tops and subalpine forests.

CULTURAL KNOWLEDGE: Luschiim described a number of uses for this tree:

> *Pashuluqw* ... that one's good. *Pashuluqw* makes very fine clothing. It [the bark] pounds to a softer material than the red-cedar. So for the families where the

husband wasn't lazy (that was just before [cloth] diaper material) ... here in Cowichan, you had to go up high to get it, so it was the man's job to go and get it. And usually the guys that would go to that extent would also have bearskin with the fur on it for bedding. So if you could walk in the house and you see yellow cedar and you see bearskins, you know  they're not lazy! ... Not every family was fortunate to have men that weren't lazy. So you see some, well, the ones that had lots of material, the cuttings get thrown away somewhere, all the cuttings. And the ones that weren't fortunate, the ladies would have to go and gather these [cuttings] ... So my great-grandparents always told me, "Don't *you* be lazy!" [laughs]. (December 16, 2010)

Pashuluqw ... nice straight grain. The wood is rot resistant. It makes good canoe paddles, whether it's everyday paddles or racing paddles or tribal journey paddles. It's good for that. Quite springy—one of the things we look for in a paddle. It's also used for carving things. It's [a] very easy material to carve and yet very strong ... You have to be careful with it, in that when it's damp or somewhat green yet, it'll check or crack very easily. You have to be careful with that. Also [it's] used for a paddle shirt, [for] dancing, [decorative paddles] three to four inches long, *pashuluqw* ... But—we just finished talking about lazy people—I was told, "If you ever make a paddle shirt, don't shame us ... by using soft woods." They named cottonwood [as a soft wood] ... (December 16, 2010)

When Luschiim took part in one of the Canoe Journeys, travelling down to Quinault, they found a yellow cedar log that, from counting the rings, was 1,200 years old (May 25, 2017).

SEASIDE JUNIPER (*Juniperus maritima*)
CUPRESSACEAE (CYPRESS FAMILY)

HUL'Q'UMI'NUM' NAME: Seaside juniper: ***ptth'une'yulhp*** (***petth'um*** "strong smelling"; see also ***puputth'in*** "skunk'); common juniper (*J. communis*; see page 138), which lies closer to the ground, is called ***slhelhuq'tsus ptth'une'yulhp***, "lying-down juniper" (September 24, 2010)

DESCRIPTION: Formerly considered to be the same as Rocky Mountain juniper (*Juniperus scopulorum*), seaside juniper is a small, often gnarled tree with shaggy bark and drooping, greenish-blue branches. The leaves are tiny and scale-like. Male and female cones are produced on separate trees. The seed cones are globular and pea-sized, bluish-green to dark blue with a waxy coating. The boughs and cones have a pungent, spicy smell, especially when

Seaside juniper (*Juniperus maritima*)

crushed or bruised. Common juniper has similar cones but has sharp needles instead of scaly leaves and is a shrub rather than a tree.

WHERE TO FIND: Seaside juniper grows along the coast of Vancouver Island and the Gulf Islands, south into the United States, on rocky headlands and edges of forests. Common juniper grows at higher elevations in the mountains on Vancouver Island.

CULTURAL KNOWLEDGE: The scent of junipers, though strong, is not considered unpleasant (see also under stink currant, page 151). Luschiim noted, for seaside juniper, "When you grab that, the branches, and kind of bruise it in your hands, a very strong aroma comes out; that's what it refers to. It's not stinky, just strong smelling" (December 7, 2010). He confirmed that it is native to this area:

> The wood itself is also very aromatic ... it's got a nice fragrance. It's been here for a long, long time. The reason why I say that is because a botanist some years back in the '60s told us that it was introduced to our shores here, and he was adamant that yes, it was introduced. But then, some time later, I was at Genoa Bay ... when you're out in the woods you recognize things, so I seen this stump and I knew it wasn't the stump of the fir tree or whatever, so I went up on the hill and to look at it [closer up]. And that stump was about that big [1.2 to 1.5 metres across—hundreds of years old]. And it's a **ptth'une'yulhp** stump. And my dad did tell me that because of the fragrance and nice smell of that, scent of that wood, when the first comers [settlers] that [came] out here, they wiped out all the ones that you could get ... Yeah. The stump must be still there. That was ... before contact anyway ... We used to have a fair amount on the Gulf Islands. Very important uses. It's used for bathing, using the branches, some of the **ha'hwthut** today, they call it **ha'hwthut**. You build a little hut, you have hot rocks and you move the hot rocks ... sweat lodge. Yeah, that's **ha'hwthut**. So that's what you use it [for]. You put it in there. (December 7, 2010)

He talked further about the sweat lodge and recollections of his mother:

Oh, yeah. They're not introduced to here. I need to say that, 'cause some people say they're introduced. No, no, we always had it here, but not in the big things that you had in some other places. I'm told they were private and your little sweat lodge is almost the size of this table, from what my mom told me. Just enough for one man. It was a private thing. She was born in 1925 and ... she used to prepare some of the things for her dad. Meaning, she'd get the branches and bring them home, she'd get some of the rocks, she'd get some material for the hut and then she was shooed away after that. She didn't see what happened. That's **ptth'une'yulhp**. (December 7, 2010)

He said the boughs of the seaside juniper were also used as a smudge or incense for purifying the house:

Yes. The same thing you do to **q'uxmin**. You do that. [NT: If somebody's sick?] Yep. You can use **q'uxmin**, as I mentioned. You can use ... grand fir, for those purposes ... Cleanse the house. There's other things, but ... (December 7, 2010)

Seaside juniper

SITKA SPRUCE (*Picea sitchensis*)

PINACEAE (PINE FAMILY)

HUL'Q'UMI'NUM' NAME: *Tth'qw'ulhp* (see *tth'uqw'*, "get poked, shot"; Luschiim, December 16, 2010); needles (of any needled tree, including scaly leaves of cedar): *ts'e'lumuts'*

DESCRIPTION: A tall, straight-growing tree with bluish-green spreading boughs and greyish, scaly bark. The sharply pointed needles spread around the twig. The pollen cones are small and reddish, and the seed cones are light brown, around 10 centimetres long, with thin, papery scales that are wavy at the tips.

WHERE TO FIND: Sitka spruce occurs in moist, shaded areas and river valley bottoms, mainly in forests on the rainy west coast of British Columbia.

CULTURAL KNOWLEDGE: Luschiim described the use of Sitka spruce as a medicine:

> So *tth'uqw'* is to get shot or poked. And that's the root word for the name *tth'qw'ulhp*, 'cause you get poked. Very, very good medicine ... My friend's father, my friend being probably 10 years older than me, 'cause he's in his 70s. His dad, on the Queen Charlottes somewhere, fell down. And there isn't too much you could do with him up there at that time. So he was just fading away and going rotten. The doctors tried, but it was going bad. So the grandpa of the day up there heard about this man that was sick. From his wound, fallen down and getting ripped open or whatever happened to him. He come and took a look, could smell him going bad. So he went up and brought back Sitka spruce pitch and brewed it up somehow and applied it on there, and it healed! Yeah ...

He also talked about the antibiotic properties of spruce; he had seen a program on television showing that spruce and devil's-club were both effective in killing several kinds of bacteria (December 16, 2010).

LODGEPOLE PINE, or
SHORE PINE *(Pinus contorta)*

PINACEAE (PINE FAMILY)

HUL'Q'UMI'NUM' NAME: *Qw'iyul'ushulhp* ("dancing-plant"; Luschiim, December 16, 2010); Kuipers (2002) gives the name **qwunulhp**, "pine," for this species

DESCRIPTION: A relatively small, upright tree with rough, greyish bark and long, dark green needles in clusters of two with dense, spreading branches. The pollen cones are pinkish and the seed cones dense, hard and woody, remaining on the branches for at least two seasons.

WHERE TO FIND: A versatile tree, growing on coastal rocky outcrops, mountain slopes and acidic peat bogs, where the crowns of the trees are often dense and rounded.

CULTURAL KNOWLEDGE: The name of this pine, meaning "dancing plant or tree," is related to a game or contest children used to play with the upturned needled twigs of this pine. If you put them on a surface and tapped them, you could make them "dance," like traditional dancing; they'd see who could

make them dance the best. "The tip of the branch. Stick it on something flat, today you could use cardboard, and make it dance! That's why they called the lodgepole pine 'dancing tree'" (Luschiim, September 24, 2010). Luschiim noted that the name is similar to that of trembling aspen, *qw'iiqw'iyul'ushulhp*, which is known for its "dancing" quivering leaves.

WHITE PINE (*Pinus monticola*)

PINACEAE (PINE FAMILY)

HUL'Q'UMI'NUM' NAME: *Ts'qe'ulhp* (Luschiim, September 24, 2010, and December 16, 2010). Note: Elmendorf (1969) gave the name *qw'əqw'əyíʔləshəlhp*, "dancing tree" (EL60), to white pine, and the Upriver Halkomelem name for white pine is similar (***qw'eyíléxelhp***).

DESCRIPTION: A tall, straight-growing tree, growing to 50 metres or higher, with long bluish-green needles in bundles of five. The distinctive seed cones are long—growing to 30 centimetres or longer—and slender when green. The cones turn light brown and their relatively flexible scales open when mature to release winged seeds, four to seven millimetres long.

WHERE TO FIND: Grows from sea level to subalpine locations. It is very rare on Vancouver Island today because it was seriously affected by the white pine blister rust (*Cronartium ribicola*), a fungus accidentally introduced from Europe in 1909, which spread across North America over succeeding decades.

CULTURAL KNOWLEDGE: Luschiim noted that Amelia Bob was the one who taught him the name for white pine. He explained about the suffix *–ulhp*, which is found in many plant names:

> The *–ulhp*, you might get asked, it's on a lot of trees. Does it mean the tree, medicine tree? What does it mean? *–ulhp*? Some years ago, back in the '60s and '70s, people doing similar work to what we're doing right now, somebody said "Yeah, yeah. It means 'tree'; it's used for medicine." *–ulhp*. Okay, then how come we've got *t'uliqw'ulhp* [name for yarrow, *Achillea millefolium*, a herbaceous plant and medicine]? *T'uliqw'ulhp* is a flower, it's used for yarrow, so it doesn't mean "tree." So the best we could make of it is that it refers to a plant used for medicine. But I'm just guessing, we're just guessing, yeah. And in Lummi, it's *–ilhch*. Variations of *–ilhch* for different places [and also found in SENĆOŦEN/Straits Salish language]. (September 24, 2010)

He talked about people travelling great distances, long ago:

> We have some Bella Bella and Bella Coola names here. There're connections there. So, it goes back pre-contact, yeah ... It sounds like, from listening to the old stories, *really* old stories, I take it that at one time we travelled a long ways, 'cause we went to the big river, waaay up that way, the Skeena. A *long* ways up that way—many days to get up there. But then there come a time when we didn't travel up there anymore, and in fact we were afraid to go too far up that way. So it means there was a shift of power, just like the old country, you know, you had France, Napoleon was great for some time, but then there was many others. So over thousands of years, there seems to be a shifting of power.

He also described how white pine was used, along with grand fir and its relatives, to cleanse traps:

So this is one of those that is mixed with that, treating your traps to keep it from rusting, de-scenting it. You also used the bark of this tree, yeah. The white pine. The white pine pitch was very good for many things. Diluting it and making it into a drink ... many different sicknesses. You could create a medicine tree. You go to a pine, you take some bark off. It's going to just cover itself with just pure pitch. [You would take off] about that much, to begin with [about 15 centimetres wide and 30 centimetres long]. The big tree that we used to go to was about that big [1.5 metres/"5 feet" across!] and the part that was open was about that big [about 60 centimetres by 60 centimetres] depending on the size of the tree.

Then they would go back again and again to the same tree to collect the pitch. But unfortunately, white pine was impacted by the blister rust fungus introduced from the east and all the big pines were killed off:

Oh yeah. Shortly after contact, a disease came to the tree, some kind of a rust disease. And they said these mountains were just *red* with that when the pine was dying. The white pine [blister rust]. It killed a lot off. [NT: There were way more pine trees in the old days?] Oh yes. So when I started logging in the early '60s, there was trees that were down in Copper Canyon, that were like, you couldn't climb up on them. White pine. [About 1.5 metres high!] Something like that. I couldn't get up on it, they were on the ground, I couldn't get up ... It really caught my eye because at that time, my dad was trying to find pine to carve with. He had trouble finding pine. [He carved] totem poles. So the few that were around, people came from, even from Saanich, over to that tree in Quamichan to get that medicine. It was the only one left around. That was easy to get to. Today, they're coming back. They're all over the mountains. (Luschiim, December 16, 2010)

DOUGLAS-FIR (*Pseudotsuga menziesii*)

PINACEAE (PINE FAMILY)

HUL'Q'UMI'NUM' NAME: Tree: **ts'sey'**; fir poles, young fir that will yield poles, or small fir logs: **tse'lumun**; fir pitch: **chumux**; little slivers in the bark of this tree: **sts'itsum'**

DESCRIPTION: Douglas-firs can live to well over 1,000 years and reach 70 to 90 metres in height. The bark, mottled reddish and tan inside, is grey on the surface and smooth in young trees, thickening and becoming deeply furrowed with age. The dark green needles, two to four centimetres long and soft to the touch, spread around the twigs in bottle-brush fashion. The pollen cones are small and reddish brown. The seed cones are 5 to 10 centimetres long, brown and woody, with distinctive narrow three-pointed bracts protruding from the cone scales. The seeds are small and winged.

WHERE TO FIND: The most common coniferous tree on the east coast of Vancouver Island, Douglas-fir ranges throughout most of the southern third of BC and is very common in Quw'utsun territory. It favours dry sites at medium to low elevations and is often associated with exposed rocky outcroppings. It regenerates quickly after fire.

CULTURAL KNOWLEDGE: Luschiim recalled that his uncle generally made his tools out of a plant that had medicinal

Old-growth Douglas-fir tree

qualities, so he used young Douglas-fir for making a gaff pole, as well as for a cane (September 24, 2010). Luschiim explained how the wood was used:

> **Ts'sey** ... many things. The small poles were used for spears and gaffs, dipnet handles. So anything [that] needed [something] small and skinny and long [three to five centimetres across]. Poling up the river. That pole I mentioned, **shtl'e'lhun'um'** [a three-pronged spear], was usually a Douglas-fir pole. Poles for the rafters and beams for the big houses—long, skinny poles—were also Douglas-fir. Some of those were rafted together up at Cowichan Lake and bundled and—nothing like the bundles you see today, you know, bundles of whatever—and drifted down the river. And it's said that you hired a certain guy to go up there. After you get it close to the river, you hired a certain man that

Douglas-fir needles and seed cones

was good at **si'win'**, so-called magic work, and you'd, he'd chant, talk to the **tse'lumun** [fir poles] ... talk to the **tse'lumun**, spiritual talk, and they'd go down the river ... they'd just go aaaall the way down there without hanging up ... Could be poles to small logs. (December 7, 2010, and December 16, 2010)

He talked more about these poles, **tse'lumun**:

They used it [the word] for when you had it, the poles here, or "Go get some **tse'lumun**." Meaning it was still standing ... And ... where I got **tse'lumun** was most likely a gravelly place. And they grew slowly. To us, the slower growing means a stronger pole. So then they grow long ... It depends on the ground ... it could be a shady rocky area and it won't grow [fast] ... So where you find **tse'lumun**, you're also going to find prince's pine [pipsissewa, **ququn'alhp**] and ... **shquli'qep'nuts** [rattlesnake plantain] and others. So the trees, like **ts'sey'**, that are growing like **tse'lumun**—to me it's an indicator for certain plants. I can see it from a whole mountain away, way across the valley: "**Tse'lumun** there. Oh, there should be good prince's pine there, and other medicines." You learn how—just like the signs on a drugstore pharmacy—you can see waaay down the next block. Same with the signs up on the mountain, "Oh yeah, there's [a certain plant]!"

Later, Luschiim talked a little more about the fir poles and their use in fishing, including over on the Fraser River:

The poles are **tse'lumun** ... That'll cover from your little 2-inch poles to, say, 10-inch poles. In fact, a log house built out of the larger poles, say a foot or so, [is] called **tse'lumunew't-hw**, "pole house" ... They also use it for the **shtl'e'lhun'um'**, which is your three-pronged spear, for down in the salt water. Could be a short pole, 10 or 15 feet. A longer pole could be, like, 18 feet and longer—[it] depends on the user. Some will [say], "Well, I'm a big man, I'm going to have a 25-foot pole!" [laughs]. But there are instances where you need a really long pole. In fact, you doubled some of them—tied them together. Because a 40-foot pole would be pretty awkward to carry around. So you'd carry two 20-foot poles or so, and put them together. So what were they for? Sturgeon, yeah! For the Fraser River. So when you got there, to your fishing place, you'd put them together and you put something on the end—some say a feather, or something of that strength—and you go along, get to the bottom, you can feel the bottom, [then] up a little bit. You're going along and you're not touching anything, then you can feel your feather touch something. And you jab it, and if you're lucky, it'll be a sturgeon. If you're not lucky, it'll be a log! [laughs]. (December 16, 2010)

The wood is also used for canes and for constructing weirs. Luschiim continued talking about the use of Douglas-fir poles:

Yep. Anything from a small walking stick, a dancer's cane or stick, yeah, so, many uses. Including **shxetl'**—weir ... Here in Cowichan we had lots of weirs at good locations. So you got a lot of these tiny **tse'lumun** ... maybe just a bit bigger than my fingers. You put them close together, tied them, and that's your ... weir that goes across the river. They're built in, say, four-foot sections or so. So when you got rid of the leaves, when the leaves are coming down, you just turn it the other way and the leaves will float away. So they're built in sections so you can get rid of the leaves quickly.

So while I'm talking about **shxetl'**, I'll mention **stth'aqwi'**. In the '50s when I started getting involved in environmental things, and one of them being salmon, I went to listen. These people were talking, saying, "You can always tell a Cowichan fish. They're different than fish—than spring salmon from other places." They said the spring salmon from Cowichan River was thin this way [across]. They were deep in width, but thin this way. And they were talking, they were wondering why. How come Cowichan fish were that way? So this was back in the '50s when I heard this. And it's common sense: after thousands of years of building these weirs, you know, the thinner ones can get through ...

So that's **tse'lumun**. Anything from, pretty well, a fat finger size to, say, 12-inch poles is **tse'lumun**. And in my early years ... [when I was] 10, 12, when I started working, one of my jobs was to carry Christmas trees out for the Christmas tree people. I saw these places that had lots of young *tse'lumun*. You know, they were small trees yet, barely bigger than my fingers, coming up. And today, [in] that same spot, some of them *tse'lumun* are maybe 15 to 18 inches. It's usually a dry place, like gravel, that grows good *tse'lumun*. So that place still has some *tse'lumun* that are usable for spear size—[about] 2.5 to 3 inches at the base. [They are] just now getting too old. So I could say that a place like that, if you see it when you're very young, and the place is also very young, it could last you your lifetime. Everything you need, you're going to get it there. (December 16, 2010)

Douglas-fir pitch, **chumux**, is also very important, and Luschiim talked about it at length:

Pitch—**chumux** is pitch. So ... your pitch, from the runny kind—the one that flows out of the tree, out of the roots or the bark—to the core of the wood that's totally soaked in pitch. And they're all useful. I'll go to the fire use first. In the past when you're walking through the old forest, you're shown the pitch sticking out of the ground, or the pitch laying there in the old wood laying on the ground. The ones sticking out of the ground are needle sharp, grey, sticking out, usually the middle. That was important to be able to recognize these, and could be still important today ... So if you're in the outdoors all the time, one day you're going to spend the night out there somewhere, or maybe just you need to make a little fire, warm up, and you watch for those, and you learn to find them even ... just a little bit sticking out; you can see that grey, sharp needle, like slivers, sticking up.

Today, it's somewhat different ... You can go to old stumps that are over a hundred years old and even though they're like that, you can see that there's where it was cut, there's a solid part there, and you kick that brown reddish rotten wood out of the way and you got your pitch in there for making fire.

One of my co-workers, her son went up Mount Heather, at north end of Cowichan Lake ... Well, this boy, he is about 20, and him and his girlfriend went for a walk and they spent the night up there. And they remembered me talking about pitch, and he looked for some and he found it. And that helped them stay warm overnight. They went down the other side of the mountain instead of coming this way! So that's one of the uses, for [starting a] fire, for that pitch, from the wood.

But it's also used to keep an ear hole open—pierced ear. You can shave that [off], the one that's really amber, brown colour. There's a lot of pitch in there. You point it really nice and smooth and ... if you want to take your earring off and leave it off for a while, you can put it in there. Pitch wood. When it's solid pitch, it's a kind of amber brown colour ... But having said that about that pitch wood in your ear, you can also clean off a feather—take everything off the feather—and poke that in there and keep it in there, if for whatever reason you want to take your earring off for a while, a while being a week or two. [It will] keep that ear hole open. (December 16, 2010)

Luschiim also talked in detail about using pitch wood for torches:

Um, pitch. Torch—pit-lamping is not new. One of the things we pit-lamped was ducks. And you used a wood torch, pitch torch. But they weren't always available, 'cause years ago we were thousands or more First Nations along the coast here. In some places the pitch was depleted, the available pitch. So what you did was you went to a fir tree, [and] you chopped a hole in the bark. I imagine the hole would look something like the hole you'd make for a springboard for falling a tree. And you stick something there, like a piece of bark or whatever you're going to use to catch the pitch running down. Then you keep collecting this pitch ... One of the things you used was rawhide. It could be anything else ... willow bark ... cedar bark. You warm up your pitch [for the torch] and you're starting to wrap your stick. And you heat up that pitch, and you pour it on there. And you wrap it some more, pour some more pitch on there. That's going to be your torch. For ducks, you go close to where they're rafted up. When the ducks are all together, we call it "rafted up." You don't have to go right up to them. You could be quite a ways, even a kilometre away, maybe more, and you light your torch. And you wait there. You've put the torch in the front of your canoe, or your stern, whichever you want. Your torch is like that [slanting out]. You can see out there. But you have a shade, which you're going to put between you and the torch so you're in the shade but you can see over to the water, and the ducks will come. And you have a six-foot or so long club, depending on your strength, for the length of your pole, and again the *tse'lumun*, and then you club the ducks that you need, when they come to investigate ... So that's how you pit-lamped ducks in the past. I didn't experience that, but I heard it from the Elders. The last one that mentioned that to me was Norman Charlie from Kuper [Penelekut]. But there's other Elders that mentioned it. Yep. That's *that* pitch! (December 16, 2010)

Pitch is also important to help start a fire when you are camping:

In the past, and even today, you're always watching for pitch [for fire starting] if you're going to be spending several nights out there. You might collect some. Today, you can have a piece of cloth or even a napkin. You can always find pitch that's running, and you collect that. Can use that for your fire. We did it on the West Coast Trail. It was pretty wet when we were over there, just pouring all the time. Six days we were there, seven days, last day before it stopped raining. And on the West Coast Trail there's lots of people and the roots of the Sitka spruce and the fir, the bark was peeled off, from continuous stepping on, so there was lots of pitch oozing out of there, so we collected that and within five minutes we had a fire going every evening ... (Luschiim, December 16, 2010)

However, cooking with pitchy wood is not desirable:

And you're cooking, outdoor fire, you make sure you don't have pitchy wood. If you're using *ts'sey'*, which is your Douglas-fir, you make sure you take the pitch off. Except for spiritual uses, and then you look for pitchy wood for spiritual cooking ... [otherwise it would flavour the food too strongly] ... You'd be looked on as a poor cook and a poor fire-maker. I could be making a fire for my mom or for my wife and, "Oh, you're a poor fire-maker. Look, you're burning pitch for the cooking!" (Luschiim, December 16, 2010)

Luschiim recognized Douglas-fir bark as an excellent fuel (although he had not heard about the practice of prying a slab of bark off a living old fir tree with wedges, as was done by some coastal peoples):

The bark of the older trees is very thick. That's a very good firewood. It burns a long time, and it glows red for a long time. If an equal amount of wood was put in there, that wood would be gone and your bark would be still glowing red, so it's hot. (December 16, 2010)

Douglas-fir wood was also prized for a variety of purposes, including long shingles:

So I know, we also used certain fir trees that were straight grained for things like shakes or wide boards ... And that has a different name than **ts'sey'**, but it's for that kind of a tree. In fact, I'm told that the name **leey'qsun** ... is "Shingle Point" village on Valdes Island, and I'm told that's the name of that other fir tree, but I don't know how it was said ... I'm told that it might be derived from that other kind of fir called "white fir," but we don't know what kind of tree that is. Its wood is as good for making shingles as cedar. [NT: The "shingles" are actually slabs of white fir wood?] Yeah, and they'd be maybe eight feet, ten feet, longer maybe. I've seen a barn that was made from that, they call it white fir, that's the local name for that ... Yeah, there's this big barn that was probably 100 feet long by 50 feet long that was all white fir [wood]. (Luschiim, December 16, 2010)

PACIFIC YEW, or
WESTERN YEW (*Taxus brevifolia*)

TAXACEAE (YEW FAMILY)

HUL'Q'UMI'NUM' NAME: ***Tuxwa'tsulhp*** ("bow tree or plant"; see ***tuxwa'ts***, "bow"; Luschiim, September 24, 2010); clam digging stick made of yew: ***sqelux***

DESCRIPTION: A small, often twisted tree, usually under 15 metres high, with spreading branches. The reddish bark is often shredded. The flat, pointed needles are dark green above and lighter beneath and spread horizontally from the twig to produce flat branches. Male and female reproductive structures are borne on different trees. The pollen cones are small and inconspicuous. The female trees produce berry-like arils, which are bright red, fleshy cups, each containing a single seed. The seeds are poisonous to humans but are eaten by birds.

Pacific yew (*Taxus brevifolia*)

WHERE TO FIND: This tree is not common; it grows in mature, moist, shaded woods at low to middle elevations, often beneath Douglas-fir, western red-cedar and/or western hemlock.

CULTURAL KNOWLEDGE: A site on the Cowichan River is called ***Ti'txwatsulhp***, meaning "little yew tree/bow tree" (Cowichan Tribes Elders' Advisory Committee, 2007). Luschiim talked about yew and the importance of its wood and of its bark and needles as a medicine:

> ***Tuxwa'tsulhp***. It's the ... yew tree. ***Tuxwa'tsulhp***. Very strong wood. It's better when treated. As I mentioned earlier about the ***q'am'*** [kelp], if the ***tuxwa'tsulhp*** is too dry, it gets too brittle, especially the heartwood. The sap is more flexible. More resin, or whatever it is ... The wood is used for tools. In some cases, you'd want it very hard. In which case you'd burn it [in a] hot fire ... But if you wanted it springy, you'd do it a different way. You'd treat it with herbs. So it's different if you want it very hard or you want it flexible. You'd treat it accordingly.
>
> And the bark is good medicine. That's some in the box up there. Some of the medicines I always keep handy. So the bark, some use the needles, some ask for needles. Both needles and bark ... Over the years, over my lifetime, I've seen ***tuxwa'tsulhp***, young trees, meaning a foot or so diameter, die off. It's my observation, when it gets too much canopy over them, they will slowly die away. So in my harvesting, I started to cut the lower branches of those trees to see if they'd perk up, and they did. So if you travel to Tzouhalem and some places where I've been,

Digging stick made from Pacific yew

you'll see some of the limbs cut off, you'll see them growing better, and where another one, not too far away, has just died away. So I've experimented with them that way.

The yew tree, *tuxwa'tsulhp*, that's the tree and the bow; that's a bow and arrow. So the bow is *tuxwa'ts* ... the actual bow, the bow and arrow. And since the bow is made from that tree, the tree is called *tuxwa'tsulhp* ... Also used for medicine, the bark, or bark and needles. Usually you'll get it from the morning or day side [of the tree] ... [where the sun hits] in the early part of the morning. (December 16, 2010)

Luschiim described in more detail how the wood was treated differently in making a bow or in making a yew-wood digging stick:

The wood is used for springy things, but the wood can become brittle. So for a bow or other springy tools or implements, you treat it with herbs. So one of the things we used was the bull kelp, the bigger ones. So you shape out your bow. Then you collect your herbs, you cut this piece of bull kelp and you stick your bow in there. You've already made your fire in the ground, and you've dug it out ready to receive that bull kelp with your bow inside it; you plug both ends, and you put it in there and leave it in there 'til it cools off, overnight, a day and a half, whatever. You treat your bow or your wood so it'll be nice and flexible and springy.

But you can also harden it by just straight burning, for tools, such as a clam-digging tool ... digging stick ... *sqelux* ...clam-digging stick. ... You could use the same word [for a root-digging stick]. *Sqelux*, *sqelux* means digging stick [The word is used for a fork now.] ... The butter clam: in the spring and early summer, especially the big ones, when you open it up, it has a kind of a clear, almost a see-though wormlike thing in there; that's called a *sqelux*. So, early summer, you get that, especially the big ones, you'll open it up and you'll see it in there. It's not curly or anything, just kinda like that. I've asked them but I never got an answer, but we call it *sqelux*. (December 16, 2010)

Luschiim repeated his observations and experiments in pruning yew trees:

The western yew, when it gets a canopy, it usually slowly dies away. So that's what I've observed over my lifetime. Some years ago, 20 years ago, maybe longer, I started wondering, "I wonder what would happen if I pruned it a little bit?" So on my trips, when I get **tuxwa'tsulhp** bark for medicine, I started to lop off some of the branches to see if that [they] grow better. And they have. So I've been doing that for a number of years now and, if you walk around here, you'll notice some of those trees with some the big branches missing and that was most likely myself or my children, getting [yew bark] ... Oh yeah, they come back. Their needles were getting very thin, you know, falling off, and after a couple of years after I removed some of the branches, they started to get full again ... If some of those branches were removed [from a dying yew tree], it might even survive to another 50 years, 100 years. I've let the Parks people know some of my experimenting and done it right in front of them to see what they'd do to me, but they watched me and let me carry on! (December 16, 2010)

Luschiim said that he had heard of the red berries of yew being used, "but I wasn't told what for" (December 16, 2010).

WESTERN RED-CEDAR *(Thuja plicata)*
CUPRESSACEAE (CYPRESS FAMILY)

HUL'Q'UMI'NUM' NAME: Cedar tree: ***xpey'*** (also "cedar canoe" and "cedar post"); cedar plank: ***xexpey'***; cedar branches: ***xpey' tsus*** ("cedar hands"); cedar that's good for a canoe: ***nuhw-lhalus*** ("canoe grain"; see ***snuhwulh***, "canoe," and ***'uyaalus***, "straight grain"); cedar house post: ***qequn*** (note: ***qequn'***, "stealing," is a different word but sometimes confused); inner cedar bark for weaving: ***sluwi'***; woven cedar-bark fibres: ***wutth'emutth't***; open-weave basket: ***kwi'kwlha'lus***; rope of cedar roots, cedar bark or willow: ***syukw'um***; cedar rope: ***kw'aythat***; ritual bathing with water and cedar boughs: ***'uqw'iwst***

DESCRIPTION: A large tree, growing to 60 metres or higher, with long, drooping branches that often turn upwards at the ends. The base of an older tree often flares outwards, with wide ribs of wood running out from the main trunk. The greyish bark is fibrous and vertically aligned. The tiny leaves are light green to brownish green and scale-like, overlapping on the flattened boughs. The tiny pollen cones are reddish. The small, elongated seed cones are green when

Western red-cedar *(Thuja plicata)*

unripe, turning brown at maturity, and usually clustered together on the outer boughs.

WHERE TO FIND: Western red-cedar grows in moist soils in forests throughout low to medium elevations on Vancouver Island and all along coastal British Columbia, as well as in the Interior wet belt.

CULTURAL KNOWLEDGE: This is one of the most important trees in the Cowichan area. Its wood, inner bark, boughs and roots have many different applications. Luschiim described the range of its uses:

> So, **xpey'** is used in past, when we were first born, we were wrapped [in cedar-bark blankets], our diapers were pounded red-cedar bark. We used them for buildings, canoes, sacred objects, masks, many kinds of tools ... And when we died, in the past we were put in a cedar coffin. So right

Western red-cedar bough and seed cones

from birth, everything in between, to death, cedar was important to us ... My mom had [a cedar bentwood box]. It was about the size of this table, bentwood box [very big; about 1.2 metres]. And probably about the same height [90 to 120 centimetres tall]. And [out of] one board. My brother's still got it. It's come apart at the seams, it's broken up. But he's got it put away. That was from her grandmother [Luschiim's great-grandmother].

Xpey' is the generic name. It has quite a few names, depending on what you're going to be using it for. And I guess those words would be gone. *Xpey'*. Um, you got straight grain ... a trunk with no branches would be very straight grain. That's what you need for making cedar shakes or wide cedar boards. They split very easily. But that's not what you're looking for to make a canoe. I've known, I know of one [canoe] that actually came apart before they used it. They were towing it home.

Culturally modified western red-cedar tree

And they got some rough weather and it split in half. And another one, same thing, a very nice grain, straight grain, no knots, this was on the US side, and the first time they put it in the water, they were paddling and it got a little bit rough and it fell apart. Yeah. (December 16, 2010).

The Quw'utsun people recognize different types of western red-cedar, as described by Luschiim, including the straight-grained type mentioned above:

Xpey'. That's the generic name, if I can say it that way, for cedar. [There are] several other names, but it's more on the spiritual side, so those aren't usually said. But I can say this: there are four main kinds of cedar. There's the swamp cedar—very dark, in fact it's almost brown inside, inside the tree, and very high resin [content]. You just scratch it inside with a tool and that resin will ooze from it. Lots of resin in there, and it's heavy.

And on the opposite end, you go to higher mountains, on the rocky bluffs, [where it's] dry. That's a very light cedar ... very light wood, would be growing on a dry rock, bouldery place, like mountainside, so, from your very light wood to your very heavy wood ... And there's two in between. That's how my dad, Simon, described it, 'cause he carved all his life, he said there's four main kinds. Certain ones are used for certain things. So somewhere in between you'll find the nice straight, big cedars, no branches for a long ways. And that's used for making planks.... They've got no branches, straight grain. They split very easy. But that's not the one you're looking for if you're going to make a canoe. [NT: So the plank cedars are not the same as the canoe cedar?] No, no, no. It's a real no-no. In fact, recently in the past, say 20 to 30 years [ago], some of our canoe builders have used that kind of cedar [straight-grained]. And one of them was being towed home ... [and] it got split ... fell apart. You [have to] find one that's got some branches—some branches, not lots. But some branches all the way, and not big branches. (Some branches are tree-sized. That's not the ones.) You find branches, say, big branches [that] would be, say, roughly 3.5 to 4 inches, preferable smaller, and that's the ones you're going to use. And the grains on a canoe cedar are somewhat intertwined. And ... after you're experienced, you get to learn what they're like … In fact, information would have been passed down from generation to generation. From my Elders, one of the places was, you travel up past Skutz [Falls], into the valley, on the Copper Canyon valley, and up there you'll find **nuhw-lhalus** ("canoe-grain") … Canoe is **snuhwulh**. And **nuhw-lhalus** is "canoe-grain cedar." It doesn't say "tree" [the "tree/plant" suffix]. So over time you get to learn these [things] … information that's passed down to you. One of the ways to tell **nuhw-lhalus** is, you pull on the bark after the tree's been down for a while and the inner bark can be quite grey, and they say, "That's going to be **nuhw-lhalus** for canoe."

And in my canoe making, I built lots of canoes in my time. And then I started to use fibreglass over some of the canoes, and the ones that are dark and got lots of that oily resin in there, it doesn't take fibreglass very well, peels off. Even though you clean it off and it gets out in the sun, that resin in there actually boils out of that wood and lets go. We tried wiping it down with a chemical called acetone, take off the surface resin, but if it's out in the sun it'll bubble up again and take it off. Um, in the recent past, and only in the recent past I guess, when there's lots of log booms, and lots of lost, broken up booms, some logs float high, and he [Simon Charlie] said that's the one you look for. But I don't know. Why is it floating high? Maybe it doesn't have much resin in there, but some prefer to use that, when they float high, for canoes. If I did that, I'd make sure there's enough knots in there to make it a canoe log. (December 16, 2010)

Luschiim reiterated on several occasions how, in making a river-going canoe, it is important to have a cedar tree with an intertwined or twisted grain to the wood (**sq'ay'tth'ulus**, "intertwined grain"), rather than a straight grain (**'uyaalus**), so that the canoe wouldn't split open if it hit a rock in the river:

The grain is such that it hangs on together. And us Cowichans living here by the river, travelling the river, we need [the grain] really tough, or **nuhw-lhalus** canoes, for going up to Cowichan Lake, Skutz Falls, Marie Canyon. A little bang and you … I made one canoe out of straight-grained cedar. I never made another canoe out of straight-grained cedar. I just tapped a little boulder in Marie Canyon and [it cracked] … I put my little brother in that canoe, because he was lighter than me and he could come the rest of the way in that canoe, the cracked canoe, yeah. So, that's a *nuhw-lhalus* canoe. Straight-grained would be *'uyaalus*. But that could go for any tree that's got straight grain, *'uyaalus*. It could even go for somebody with very good eyes to see things, *'uyaalus*. So it's good eyes, good grain, it would do for both of them. (December 16, 2010)

He noted that certain places were customarily cared for by families or groups, or a given community, so information about places that produced good canoe cedar trees would be passed down through the generations within a family: "It's most likely passed down to you, like that information was passed down to me … Yeah. I might grab any cedar from close by, but for a special canoe I might go up there [to a special place]" (December 7, 2010).

He then recalled what he was told by one of his Elders, Ben Canute, from when Ben was very young, in the early 1900s:

He seen two canoes down at the river. He observed them many times. So one time he asked, "Those canoes down at the river—below Stone Church, near Green Point—how come they're there?" And the old man of that day—this was around 1912—[answered], "Oh, they were built about 100 years ago. There was 40 of them made all at once." Those two canoes: one was 65 feet and one was 70 feet. That's what I was told. One was 12 feet wide and the other was 14 feet wide. And they were made at Qul'i'lum' [now Dougan Lake], that's where they were made at, all 40 of them. So, about 100 years ago—about 100 years from 1912, thereabouts—would have been 1800, 1812, somewhere there. And there is a record of a war, where we went and warred on somebody at that time … in retaliation for a raid over here. So, I'm guessing, it might have been at that time … (December 7, 2007)

Cedar was also used for house posts and beams, and these were also very carefully chosen:

They're very important to have as the main posts, the main **qequn**, house posts ... And some of our speakers, we speak to the post as we're coming in, using the female and the male name as we're coming in. I did mention before, I touched on part of it here, but there's four main kinds of cedars, but even within those four there's other categories, if I can put it that way. So the ones from the swamp, with dark wood, are very high in resin, very dark colour ... (Luschiim, December 16, 2010)

Luschiim also described the importance of cedar boughs:

Cedar branches, the same as the tree, **xpey'** ... **xpey' tsus** ... **tsus** refers to the hands ... You use the **xpey' tsus** for certain things, one of them being the masked dance. The masked dancers hold the **xpey' tsus** on one side and the shells on the other side. So in case you ... share this, here is one: [The person who] goes to get the **xpey' tsus** for the masked dances, [they] try and get the firmer ones. Some are firm where the branches are like that [straight out] to the tip; the other ones are like this [hanging]. You want to get the firm ones. **Xpey' tsus** would be used for bathing up in the mountains ... You scrub with it like a washcloth. But when you finished with it, you finished your [bathing] ... you put it up on the trees somewhere, where the wind is going to brush it off [to remove any negative influences]. (December 16, 2010)

The boughs were also used to make sturdy openwork clam baskets:

My great-grandfather Luschiim used to take about half an hour to make a basket out of branches. He'd accompany my mom, Violet, and my granny Celestine, whenever they went out in the salt water. His home was at Lhumlhumuluts', about half a mile up from Cowichan Bay. So they went out often. He'd always accompany them when they went out. And they stopped for clams; my granny would be digging, my mom would be picking—she was quite young, about 5 or 10 years old—she'd be gathering the clams, and Luschiim, my great-grandfather, went up on the beach into the trees, and he'd take these cedar branches and split them and he'd make a clam basket, just a small clam basket. So by the time they were finished digging, he was there with the basket and put them in there, swish-washing it. So every time they went out, he'd make a new basket for his daughter, my granny. So the branches were used, the roots were used for making baskets [NT: In making the clam basket, do you twine the split branches with the roots, or what do you use to weave with?] It'd be ... just the branches, I think, but you'd weave it in such a way that it's not a straight weave, 'cause a straight weave would move. So every spot was, to weave it is to **wutth'alus**. Any straight weaving, wool or basket making, but the **lhun'ut**, you tie it on each other like that, so it's not going to move. (Luschiim, December 16, 2010)

Different weaving techniques are used for different baskets:

So there's names for every kind of a weave, most of them are lost. I just know the two: **wutth'alus**, just a straight weave, or you could go over two, under two, over two, or whatever, there's different ways of doing it; and the **lhun'ut** is to put that special weave in there that's going to keep it from moving. An ordinary basket, straight weaving, **swe'wutth'**, would always have the top **slhulhin'**, that's the same word, meaning ... it's done that way to hold it together. A bigger basket would have **slhulhin'** every so often to keep everything from moving around. So you also used roots, the same way ... any part of it could be used for rope—your branches, your roots, your bark—for rope. Could be anything from a rope the size of your thigh to a thread-like twine, yep. (December 16, 2010)

There's another word that goes for bark, or reeds, yeah, I lost the word. But what it is [made from], the bark or reeds, is not all the same. Some will have the ability to withstand, I'll say rigorous use, or they may fall apart easily. So the places where the good cedar or reeds, is kind of a family-guarded knowledge and passed on from grandmother to granddaughter, type of thing. The word for "brittle" is **thuphwum**, so if you got some cedar or reeds and your work fell apart very soon after you start using it, then you don't go there again. (Luschiim, December 16, 2010)

Luschiim explained that **xpey' tsus** refers to cedar branches, or boughs: "**tsus** is to do with the hand; it's the hands of the cedar." He also noted that they might add the suffix **–ulhp** when referring to the cedar tree, whereas for the Lummi, Nooksack and in other areas they would use the suffix **–ilhch**.

He said that cedar boughs were and are used for bathing and other spiritual practices, and some people used them as a scent, although his family used **p-th'une'yulhp** [juniper]. Cedar branches used in a spiritual way were treated with extreme care and respect, although some do not follow the protocols well today:

The **xpey' tsus** is used for, like when you **kw'aythut** or you go for your mountain bath. Like, you would use a washcloth, brush off with it. Also used for masked dancers, use on their left hand for their work, brush off for their work. But whenever you use cedar branches you're asking, you're doing spiritual things with that branch, and you have to be respectful to that tree, to them branches. You let it know what you're going to use it for, apologize for having to break some of the branches off. From then on, you're responsible for the well-being of that branch or branches. Once you've taken it off the tree, you never let it touch the ground or the floor. If you have to put it down, you shield it or put down some kind of a protection thing underneath it, even if you have to take your coat off to lay it down. Preferable for you hang it up somewhere. And then you respectfully take it up, after you've finished with it, up somewhere up in the forest and up into the mountain. If it's been used, you make sure it hangs up somewhere. You're going to let the wind blow through it, to cleanse off anything that's on there.

And then, sometimes, that particular branch or branches is passed on to someone else. Whoever touches that branch is now responsible for the well-being of that particular branch. If you've got to take some of it off, to make it a smaller branch, you're still responsible for those little pieces, that it is all taken care of in a good way. That's something that's missing, 'cause I see cedar branches laying all over the place, when they decorate a hall, after, it's just laying down. That's not the proper way. For spiritual uses, I see the same thing ... I'm hoping that you'll be able to share, not to impose on them, but to share that knowledge with other people. Then it's, once they know it's their responsibility to follow through. (December 16, 2010)

Luschiim observed that recently cedar trees have been attacked on the inside of the tree by the powder worm, which weakens them. He said it is hard to tell whether the trees growing up now will be as plentiful as in the past (December 16, 2010). He talked further about the changes in cedar he has seen:

I went to Salt Spring on a referral, and we looked at one of the places that was recently logged. The logs were still laying there. There's one log that was laying there, we figured it to be 250 years old and it was this big [about 1.5 metres across]. And there was another one further up the road and we counted the rings, and it was tight grained—old growth ... about 800 years old. And they were same height. So second growth ... how come there's so much difference in the diameter of the tree? One is really fast growing, which is that 250-year-old at breast diameter. In the past, the cedars that survived were under the canopy of the forest. A forest, say, 80 years old—I'm just going by what I see out there today—I see 80-year-old second-growth fir, with lots of cedar underneath, a second set of trees coming up ... And they're going to grow slow because of the canopy over them. And that's tight grain that we look for ... Today, when the forest is logged off, the cedar grow up with the rest, so they're growing fast, because they've got lots of sunlight. I don't know if that's why they're getting attacked by the powder worm or not, but that's something to think about.

Merv Wilkinson [at Wildwood, Yellow Point] ... explained, they were kind of experimenting. It seemed like the cedar under the canopy did not get attacked as badly by powder worm, and that the ones out in the open, such as the second growth setting out in the open, they were getting attacked by the powder worm. (December 16, 2010)

WESTERN HEMLOCK (*Tsuga heterophylla*) and MOUNTAIN HEMLOCK (*Tsuga mertensiana*)

PINACEAE (PINE FAMILY)

HUL'Q'UMI'NUM' NAME: Western hemlock: ***thq'iinlhp***; mountain hemlock: ***tsa'luqw thq'iinlhp*** (literally "high elevation hemlock")

DESCRIPTION: Tall, straight-growing trees, up to 60 metres tall; the younger trees have drooping tops. The branches tend to hang downwards. The bark is rough and reddish brown. The needles are short, flat and blunt tipped, of varying lengths but generally short and irregularly spaced throughout the branches, and whitish underneath, giving the boughs a lacy look. The pollen cones are numerous and small, and the seed cones are relatively small as well, usually 1.5 to 2 centimetres long. Young cones are greenish or purplish,

Western hemlock (*Tsuga heterophylla*)

but turn light brown as they mature. The related and quite similar mountain hemlock inhabits upper elevation forests with yellow-cedar and amabilis fir. It has longer cones that are deep purple when young, and its boughs are more brushy-looking than those of western hemlock.

WHERE TO FIND: Western hemlock grows in moist, often shady sites. It is more common on the wetter west coast of Vancouver Island, where it grows in the shade of other trees. It occurs mostly at low to medium elevations, whereas the closely related mountain hemlock is generally found in the cool, damp forests of higher elevations.

CULTURAL KNOWLEDGE: Luschiim recognized the two related species of hemlocks and noted that the mountain hemlock has bigger cones. He noted that hemlock wood is flexible, and he had heard that some part of the hemlock [probably the knots, as described by Ditidaht plant expert John Thomas (Turner et al., 1983)] was used in the past to make fish hooks. They were steamed and then bent into hooks (Luschiim, December 7, 2010). He also talked about the tree's spiritual importance: "This is kinda lost, but we at one time used it—this or cedar—for the masked dance. So I don't know why it's faded away, the use of it. It's used for a spiritual headdress. But that's not done anymore ... I remember when I was a little boy [four to five years old], the old people talking about it" (December 7, 2010). Hemlock boughs were used for spiritual bathing by the Nlaka'pamux, according to Annie York, whom Luschiim had heard of. The name in Nlaka'pamux, *xwiqwestnelh* ("scrubber tree"), is close to the Hul'q'umi'num' word for ceremonial scrubbing, *xwiqwit*. "Same root, it's really close," Luschiim noted (December 7, 2010).

Mountain hemlock (*Tsuga mertensiana*)

Broadleaf maple (*Acer macrophyllum*)

BROAD-LEAVED
TREES

VINE MAPLE (*Acer circinatum*)
ACERACEAE (MAPLE FAMILY)

HUL'Q'UMI'NUM' NAME:
P'ene'ulhp (also used by some for Douglas maple, also known as Rocky Mountain maple; Luschiim, September 24, 2010, and January 23, 2011)

DESCRIPTION: A small, multi-stemmed deciduous tree, up to eight metres tall, with flexible branches and pale green to reddish bark. The leaves are opposite, typically maple-like, with seven to nine lobes. They are bright green in spring and summer, turning bright red in autumn. The flowers are small, borne in few-flowered clusters, with white petals and red sepals, and the seeds are winged and usually paired, attached together at or near 180 degrees.

WHERE TO FIND: Moist forests and streamsides; found in only a few locations on Vancouver Island, such as the vicinity of Robertson River near Lake Cowichan. More common on the Lower Mainland of BC.

CULTURAL KNOWLEDGE: Luschiim doesn't know the underlying meaning of the name *p'ene'ulhp*, except as the name of the tree. He noted that vine maple "likes its toes pretty damp ... It is springy, so it's used for items you want [to be] springy ... There's several things that require something springy ... and one of them is ... a hanging baby cradle ... you have a cord attached to it ... [to] your foot ... and mothers moving it ... Hanging, eh?... When you're working, every once in a while you move it, if you need to ... yeah. And also ... the *p'ene'ulhp* and *sits'ulhp* [Douglas maple; see page 87] ... [are] used for ... paddles or clubs" (April 16, 2015, and January 23, 2011). (The wood is also used for bows and snowshoes by mainland Salishan-speaking peoples.)

DOUGLAS MAPLE, or ROCKY MOUNTAIN MAPLE (*Acer glabrum*)

ACERACEAE (MAPLE FAMILY)

HUL'Q'UMI'NUM' NAME: *Sits'ulhp* or *siits'ulhp* (see *siits'ul*, "to dry out"; Luschiim, September 24, 2010, and January 23, 2011—the name was told to Luschiim by Rita Joe, née Point Johnny)

DESCRIPTION: Small, often bushy tree up to 10 metres tall, with smooth reddish to greyish-purple bark. Leaves opposite, smooth, three- to five-lobed and coarsely toothed, turning crimson in the fall. Yellowish-green flowers are borne in small, flat-topped clusters; winged fruits in pairs, attached together in a "V" shape.

WHERE TO FIND: Open woods, shorelines and rocky slopes; sporadic, mainly along the coast and along lakeshores on south and central Vancouver Island.

CULTURAL KNOWLEDGE: Luschiim explained the meaning of the name for this maple: "See [*sits'ulhp*] ... the word *siits'ul* ... *Siits'ul* is something that's dried up, probably dried up and shrunk." He gave the example of blackcaps that have been exposed to hot, dry weather and they dry up: "That berry when it's dry it's *siitsul*." The term can also be applied to people who become skinny and wrinkled as they age. He explained why Douglas maple is called by that name:

> *Sits'ulhp* likes its toes pretty dry—the roots ... In some cases, you'll see them right alongside creeks in the river ... and people say "I thought you said it likes its toes dry" ... On the river, the ground would be like that [sloping], that part of the terrain is well drained ... So people think it's because of the water, but no, it's because it's well drained, it's why they're growing there. (April 16, 2015)

Luschiim noted that both vine maple and Douglas maple leaves tend to get bright red growths on the underside of the leaves (from eriophyid mites) that look like spatters of blood: "Yes. So I don't know if it's one or both of them, they have little red growths under the leaves and it was said to be the hummingbird's menstrual pad" (April 16, 2015).

Douglas maple wood is used for carving the miniature paddles used to decorate regalia for dancers' club jackets, and mainland Coast Salish use the wood for snowshoes and bows. In other places, people use the inner bark for weaving baskets and trays.

BROADLEAF MAPLE, or
BIGLEAF MAPLE (*Acer macrophyllum*)

ACERACEAE (MAPLE FAMILY)

HUL'Q'UMI'NUM' NAME: *Q'umun'ulhp* or *ts'alhulhp*

DESCRIPTION: Large, spreading deciduous tree, growing up to 30 metres tall. Young bark is smooth and greenish, becoming greyish-brown and furrowed with age. The leaves are very large: some up to 30 centimetres wide, deeply five-lobed, with sharp-pointed tips. They are opposite and bright green, turning golden yellow in the fall. The flowers are small, yellowish and numerous, borne in drooping clusters and blooming in spring, before the leaves have matured. The fruits are coarsely hairy, with smooth wings in angled pairs.

WHERE TO FIND: Very common in Quw'utsun territory in moist to moderate woods, along river valleys and on open slopes from low elevations to mountain slopes.

CULTURAL KNOWLEDGE: Straight-grained, suitable for paddle making. Ts'alha'um (Tzart-Iam IR 5) is called "place of maple leaves," after the bigleaf maple trees there.

The leaves of this maple were used to clean and flavour a freshly killed deer; Luschiim explained that when a deer was killed in the old days, they would clean it and fill the cavity with maple leaves and/or fireweed; this helped to soak up the blood and to start the process of seasoning the meat, giving it a really sweet taste (September 24, 2010).

Luschiim explained the process in more detail on another occasion:

> My family, we use maple leaves ... for kind of a seasoning. So when you first cut your deer, as soon as you put your deer down, you rush over there, you clean it, gut it, and if there's some blood in there ... you gather a bunch of this [bracken and maple leaves], wipe out the inside and then ... we packed the deer ...stuff the inside, the ... stomach cavity of the deer with that. And it's just like seasoning. You know, at home today, you add a whole bunch of seasoning for your cooking? Well that was our way down here where there's maple. So ... the ones that I use [for seasoning]: maple, down here where there's maple. Later on, when there's fireweed, we use fireweed. And also the bracken fern. So right away, it seasons the meat. And when I slowed down in hunting, the boys were giving me deer, but there was something missing. It took me a while to find out what was missing, 'cause they weren't seasoning it, the way I'm used to it ... the ones I used was maple ... That really flavours it good (December 7, 2010, and May 25, 2017).

Maple wood was formerly used to make paddles for canoes, as well as spindle whorls for spinning wool.

RED ALDER (*Alnus rubra*)
and SITKA ALDER (*Alnus viridis* ssp. *sinuata*)

BETULACEAE (BIRCH FAMILY)

HUL'Q'UMI'NUM' NAME: Red alder tree: **kwulala'ulhp** (see **kwulula'alus**, "orange-coloured"; Luschiim, June 15, 2007, September 24, 2010, and December 16, 2010); liquid sap of alder, maple and other trees: **sx̱ém'eth**, **sxe'muth** (edible liquid sap "when it's runny"; Luschiim, June 15, 2007); edible cambium (inner bark) of alder or maple when it gets gelatinous or pulpy: **sxe'muth-us**, **sxa'muth-'us**, **sxá'meth-es** (Luschiim, June 15, 2007, and December 16, 2010); Sitka alder: **kwikwalala'ulhp** (literally "little red alder"; Luschiim, September 24, 2010)

DESCRIPTION: Red alder is a deciduous tree growing up to 25 metres tall. The bark is often whitish or light grey due to lichens; the inner bark turns bright

Red alder (*Alnus rubra*)

orange with exposure to air. The leaves are short-stalked, elliptical and alternate, coarsely toothed around the edges. They turn brownish in the fall. The male (pollen-producing) catkins, hanging in clusters, mature before the leaves expand in the spring; the female (seed-producing) flowers are small egg-shaped cones, at first green, turning brown and woody as the seeds ripen. Sitka alder is smaller and bushier, with leaves that are more finely serrated around the edges.

WHERE TO FIND: Red alder is a common tree of moist woodlands, floodplains and newly cleared areas, from sea level to mountainsides, occurring throughout Quw'utsun territory. Sitka alder is found on moist slopes, streambanks, avalanche tracks, bogs and fens throughout Vancouver Island.

CULTURAL KNOWLEDGE: The runny sap of red alder is edible and sweet. Luschiim explained: "***Sxe'muth*** is the sap, the runny sap. And for the ones you eat, maple or alder [sap], you peel [the bark] when the sap is running, you

Sitka alder (*Alnus viridis* ssp. *sinuata*)

scratch it [the soft tissue] off and you eat that. It's really good." In the next stage, the inner bark becomes pulpy and gelatinous. "And that's **sxe'muth-us** ... that's pliable, it's sort of spongy. On certain trees, any time of the year, you can survive for a while on **sxe'muth-us**, on certain trees. That was important to know" (June 15, 2007, December 16, 2010, and April 16, 2015). Luschiim's great-grandfather and namesake, also named Luschiim, taught him about the importance of alder cambium as an emergency food when he was only about four years old (September 24, 2010).

Luschiim described the edible qualities of the sap and inner bark on several occasions:

> You can eat it anytime they're there. The sap is there, probably early May, maybe even before that, 'cause it starts to get kind of bitter, maybe late summer. So alder sap, or any sap, is **sxe'muth**. And ... the cambium is **sxa'muth-us** ... You can also eat that too. So the sap is the runny stuff and you scratch that—some of it will almost be getting to be **sxa'muth-us**— it'll be kind of a film. But the sap and that little bit of a film is your **sxa'muth**. You can eat that, you can get a whole clump and you can bring it home. And you can tell when the kids have been eating that, 'cause they'll be orange! ... The orange is specific to alder, but the **sxa'muth**, you can get it from maple. Some will get it from **tsuw'nulhp**, cottonwood, but not here in Cowichan. And the cambium—the **sxa'muth**—is kind of spongy ... Depending on the tree, something like that thick [about one millimetre]. Outer bark, and then the cambium, and [then] the wood, if it's not sap time. (December 16, 2010)

The bark of both red alder and Sitka alder can be boiled in water to make a red dye, used for colouring cedar bark and other materials. When asked if he had heard of using flowering dogwood bark as a dye, Luschiim said he had never heard of that. "Here for something dark we'd use **kwulala'ulhp** [alder] or ... **t'a'hw** [grand fir] ... for dark, almost black, dark grey. How dark, usually you add more or less bark in there to give different shades" (January 23, 2011).

Alder wood is a valued fuel for smoking salmon and cooking and is also used for carving.

ARBUTUS, or
PACIFIC MADRONE *(Arbutus menziesii)*

ERICACEAE (HEATHER FAMILY)

HUL'Q'UMI'NUM' NAME: *Qaanlhp* (Luschiim, January 23, 2011)

DESCRIPTION: Canada's only native broad-leaved evergreen tree, arbutus has a stout, sometimes gnarled trunk and smooth branches. It grows up to 30 metres high. Its bark is unique—bright reddish brown and smooth, mostly peeling off and shedding every year. The leaves are oval, bright green and shiny on top, turning yellow and falling in midsummer. The small whitish urn-shaped flowers grow in dense clusters and mature into pea-sized orange-red berries, which are edible but dry and seedy.

WHERE TO FIND: Arbutus grows on dry, rocky open sites, especially along the coast, and in open woods, often with Douglas-fir.

CULTURAL KNOWLEDGE: Arbutus wood was used to make cooking sticks for bannock. Luschiim explained:

We had a visitor, when we're down the Narrows, down in the Samson Narrows area … where we needed to cook more than the tins that we had to cook with … so … we heated up a … branch of an arbutus, green arbutus. You heat it up hot, as hot as you can, without burning it too much. You wrap your dough on there. Then you cook it, like you do a barbecue salmon. Like a bannock. Yeah, so that's one way we use it. [If] we don't do that then the inside won't cook as evenly. (January 23, 2011)

He noted, however, that the W̱SÁNEĆ people do not break branches from arbutus: "If you're Saanich [W̱SÁNEĆ], you don't break a branch off that. It's a sacred tree down there. It's part of their flood story …" (January 23, 2011).
Luschiim added:

So here in Cowichan, one of the teachings [is] not to allow a pregnant woman … to go close to that [arbutus], because of the peeling bark. Is **xe'xe'** to go close to that. It's used for several medicines. Number one, the bark is peeling off—you get that bark peeling off the tree—the dry ones. You get that and you make a mild tea out of it, to wash the baby's mouth with, when they get … thrush. A gum and mouth problem. So you wash the mouth out. The green bark, it's one of the ingredients of many different kinds of medicines. (January 23, 2011)

The bark and leaves are also used to treat wounds. "I did … experience, where a wounded deer, the other deer, not the deer that got shot, but the other, chewed **qaanlhp** leaves … along with other kinds of tree needles, put it on the wound of the deer. That's how we learned about some of the medicines. I was the one following the deer, that's why I know. The deer got away in the end" (Luschiim, January 23, 2011). Luschiim had also heard of people using the bark or leaves to treat coughs or colds.

Arbutus berries were not eaten, although they are edible. "Some of the things that are edible weren't readily eaten, because we have plenty of other food around here" (Luschiim, January 23, 2011).

PACIFIC FLOWERING DOGWOOD (*Cornus nuttallii*)

CORNACEAE (DOGWOOD FAMILY)

HUL'Q'UMI'NUM' NAME: *Kwi'txulhp*

DESCRIPTION: The floral emblem for British Columbia, Pacific dogwood is a medium-sized tree with oval-shaped, pointed and smooth-edged leaves in opposite pairs and smooth, greyish bark. The large, white flowers are actually dense clusters of flowers surrounded by petal-like bracts. The fruits, also in dense clusters, are small, orange-red skinned and fleshy, each with a hard seed.

WHERE TO FIND: This tree is not as common as formerly, growing sporadically in open woodlands and road edges. It is most prominent when flowering in late spring, and in the fall when its leaves turn reddish to peach-coloured.

CULTURAL KNOWLEDGE: Luschiim noted how dogwood could be identified, even in winter without its leaves: "One of the things I learned very early was how to recognize trees any time of the year. Like right now [January], there are no leaves, but you can tell the dogwood, *kwi'txulhp*, the way the branches grow. The branches grow wavy [like waves]. If you can't tell any other way, that's one of the ways [dipping between each node]" (January 23, 2011).

The wood is valuable for making tools, such as a club for hitting a fro, used for splitting cedar shakes:

> When it's dry it's a very, very hard wood. One of the things we used it for—my dad, myself and many others—was a club for the fro, for splitting shakes. So you have this club that you hit the fro with. Your fro wears out quick. If you're not lazy you can prepare this for yourself months in advance, a year in advance. You get your club for the fro. That's one of the things you can do, if you want it to last, you go right to the very bottom [of the dogwood tree]. The grains are *sq'ay'tth'ulus* ... It means they're intertwined, for any kind of tree. If you're in a hurry, you can go look for one that's already dead. And sometimes the bottom will be already starting to rot, even before it's visible that it's dying. So the next thing to do is get a tough part; get one where the branches are coming out ... Right where the branch is, you fix your club, so that's the part you're going to use, with that branch sticking out ... (Luschiim, January 23, 2011)

Dogwood wood is also used to make tool handles, for picks, shovels, "whatever you want to fix, find a piece that's good." The wood can be used as a fuel, but Luschiim said, "I didn't particularly like it for smoking salmon. Maybe I'm just more used to the alder taste of the smoked salmon" (January 23, 2011).

Luschiim also noted that the bark is a major ingredient in many different medicines (January 23, 2011).

CASCARA (*Rhamnus purshiana,*
syn. *Frangula purshiana*)
RHAMNACEAE (BUCKTHORN FAMILY)

HUL'Q'UMI'NUM' NAME: *Q'ey'xulhp*

DESCRIPTION: Cascara is a relatively small deciduous tree, up to about 10 metres high, with smooth, greyish bark. The leaves are similar to alder leaves, with prominent veins running in parallel lines out from the midvein, but the edges of cascara leaves are smooth, not toothed. The winter leaf buds do not close up tightly; the tiny embryonic leaves spread outwards. The small greenish flowers are borne in small clusters and ripen into small, shiny black berries, not generally eaten by people.

WHERE TO FIND: Cascara grows in moist, open woods and is relatively common on southern Vancouver Island, from low to medium elevations.

CULTURAL KNOWLEDGE: Cascara wood had specific properties that made it useful, as explained by Luschiim:

> Certain woods, certain trees, certain branches have a different, I guess
> you might say, flexibility. For my dad, Simon Charlie, who made

many tools for his ... carving, different types of tools, this was one of his favourite handles, because of what he called "just right amount of spring." If it's too stiff, your handle, your hands would get stiff ... after days and days of chopping. Just the right amount of flex ... So that's one of the wood uses for that one [cascara]. (January 23, 2011)

Luschiim also knew of the use of cascara bark as a laxative medicine:

I think it's well known that it is a laxative [the bark]. For some it works quicker than for other people ... some can't even go close to it, if they're freshly cut, it'll work on them ... Also, it's very good for external injuries ... washing cuts. Mix that with **kwulala'ulhp**, red alder, and most likely **t'ulum** [cherry] bark ... wild cherry, for a wash. (January 23, 2011)

Cascara berries were not eaten, according to Luschiim, but he commented, "I'm not sure, but I believe that was one of the dyes, the berries" (January 23, 2011).

During the 1920s to 1950s and '60s, many people harvested wild cascara bark to make a little money. Pharmaceutical companies would buy the bark for a certain amount per pound. Luschiim recalled:

Probably 1946. I was about four years old. I don't recall how much I was getting paid. I do remember, later on, it was about 18 cents, or something like that, a pound. We peeled them. We squished them up till we get as much as we can in a small container. Take it to town, sell it. The dealer, the collector, was the second-hand store. So he collected them, passed them on to whatever [pharmaceutical company]. So the second-hand store, that's also where they used to buy bottles.

Continuing that story, about when I was about six years old, with the hunters of the day. That's when I started shooting. That's when I decided to buy a gun. So right away I started selling bottles, collecting bottles and selling cascara bark. So by the time I was nine years old I had enough money to buy a gun. The gun was 19 dollars. I still had it when my daughter was born in 1964 ... so it served me well! (January 23, 2011)

PACIFIC CRABAPPLE *(Malus fusca)*

ROSACEAE (ROSE FAMILY)

HUL'Q'UMI'NUM' NAME: Crabapple fruit: **qwa'up** (some people also use this term to refer to the whole tree; Luschiim, January 23, 2011); crabapple tree: **qwa'upulhp** (Luschiim, January 23, 2011); domesticated apples: **'apuls**

DESCRIPTION: Pacific crabapple is a relatively small deciduous tree, resembling a domesticated apple, with rough, greyish bark. The leaves are oval, pointed and finely toothed on the edges, often with a coarse tooth or pointed lobe along one or both margins. The flowers are white to pinkish, in rounded clusters. The fruits are clustered and long stemmed, small and elongate, greenish red or yellowish when ripe, in late summer or fall.

WHERE TO FIND: Pacific crabapple grows, sometimes in thickets, in moist ground, along creeks, lakeshores and beaches. It often occurs in river estuaries and is common on Vancouver Island.

Pacific crabapple fruits

Pacific crabapple flowers

CULTURAL KNOWLEDGE: Luschiim noted that some people use the name **qwa'up** for the tree as well as for the fruit: "Some of them just stop at **qwa'up**; they mean the tree ... because it will make a difference with the fruit.

If you mean the tree, you put **-ulhp** on it. You could just name the tree **qwa'up**" (January 23, 2011). He said that crabapples grow down in the lower Cowichan valley. "In the recent past it would grow in areas where there are fields, along the fence line here, but now they're growing all over the place ..." (January 23, 2011).

Wild crabapples are harvested late in the year. But they were no longer being harvested when Luschiim was growing up. He commented, "I was coming on too late to do any harvesting of that ... There's other information out on how to store it. I don't know how you store it. I don't know, I just heard there was some information" (January 23, 2011).

Luschiim's main memories of crabapples are from hunting grouse:

> As a growing youngster and having a gun, it was one of my responsibilities to bring food for on the table for the family. So this was one of the things I had to know, was where were all the crabapple trees. And then in the evening, I'd listen for the grouse to come ... we call them "willow grouse." You hear them when they come to land and the wings are hitting the branches they go to sit on. The branches are hitting their wings ... So I caught many grouse that way. And also, another way: if you just listen, or maybe you are trying to wait for early evening, but there's no luck. Maybe you heard a pheasant or even one of those grouse and it's too dark to shoot, so you have to wait and wait and wait ... After dark, when it's really dark, you go sneaking over there. You use the lighter sky, and you can see outline of the bird ... And you can get your meal. I got many grouse that way. So that's one of their favourite places to roost, the crabapple tree." (January 23, 2011)

Crabapple wood is hard and used similarly to the wood of dogwood: "It's a very hard wood. That one's even more intertwined. It's a very strong wood. It's difficult to get a straight piece for, like, a shovel handle. It's twisted. It's good for fuel. Any really hard wood burns well" (Luschiim, January 23, 2011).

Crabapple bark is known as a good medicine: "Yes, it's one of the ones that we mix with the ones we're naming [cascara, dogwood, bitter cherry]. It's used now for a variation of medicine for cancer ... some use it for an eye wash but I never have" (Luschiim, January 23, 2011).

BLACK COTTONWOOD

(Populus balsamifera ssp. *trichocarpa)*

SALICACEAE (WILLOW FAMILY)

Black cottonwood with white-downy
catkins visible

HUL'Q'UMI'NUM' NAME:
Tsuw'nulhp; thick bark of
cottonwood: **qwoonulhp**

DESCRIPTION: Cottonwood
is a deciduous tree growing
up to 50 metres tall. The
bark is smooth and greenish
in young trees, deeply
furrowed and greyish in
older trees. The buds are
resinous, sweet-smelling.
The heart- to wedge-
shaped leaf blades are long
stemmed and finely toothed
around the margins. Male
and female catkins are
produced on separate trees.
Soft, downy "cotton" carries the seeds on the wind in late spring.

WHERE TO FIND: Cottonwood grows in damp soil, often along the edges of
rivers or lakes, and is common in the Cowichan area.

CULTURAL KNOWLEDGE: Luschiim noted that **tsuw'nulhp** commonly grows
by the river and is named for this property: "So a place by the river, or
someplace down below somewhere, below, is **tetsuw'**—"down below." And
that's the root for that word, **tsuw'nulhp**. So it's a tree that likes to be close to
the water, or down below" (January 23, 2011). He recalled:

> As a youngster growing up, one of the things I remember being
> cautioned on [was] which trees you can climb or play on, and what not
> to. This [cottonwood] is one of the ones you don't climb. If you do, be

very careful because the branches break very easy. That's one of the things we have to bring back, is the simple teachings like that. So that's definitely one of the ones that can break very easily. If you have to climb, keep your hands and your feet right next to the tree. If you go out a little bit, even, you can break the branches ... (January 23, 2011)

Luschiim also noted the strong smell of cottonwood and talked about logging cottonwood for pulp when he was a boy:

One of the things I learned very early is it's a very, I don't know if you can call it stink or ... its smell is not pleasant. If you gotta carry it or chop it or something, your clothes, the smell will never come off! ... The dampness, the wet part, yeah. When I was probably seven, eight years old, they were harvesting *tsuw'nulhp* all along the river, to sell it to the pulp industry. So I was down there, helping, more likely getting in the way, I guess! So we split it like fence posts, that shape, and piled it, and made stacks and stacks and stacks all over the place. The piles were there part of the summer to dry out somewhat. And then later, they come along with a horse and wagon to load them up onto the wagons to bring them to the road so the trucks could come and pick them up. So here, they were harvested in the mid '40s, maybe early '50s. (January 23, 2011)

When asked about making a dugout canoe from a cottonwood log, Luschiim responded:

Yes, some people like to experiment and, say, a, what kind of canoe would this make? Yes, you can make it out of cottonwood. We used to be fortunate in that we had a lot of big cedar. You can spend a lot of time and energy building a canoe, and here we had the cedar, so only as a kind of joke or something would you [make a cottonwood canoe]. So with no cedar quickly available here, there's a few who were actually on the watch for a good big one, you know, five feet through or whatever. But by the time they're five feet through, most of them are rotten in the middle ... (January 23, 2011)

The sticky buds of cottonwood in the spring can be used to make a fragrant salve or face cream, as a cosmetic:

> Okay, the buds. The buds were used for mixing with the *'anuw'* [or *ano'*] ... deer fat, rendered. It [the fat] comes out pure white, just like this [paper], pure white. And, for cosmetics, if you wanted to make it smell sweet, smell good, then you added some of the cottonwood buds in there. I've never done it. I imagine it kind of changes the colour a little bit. (Luschiim, January 23, 2011, and December 7, 2010)

The red hair roots of cottonwood were used to make a shampoo, as well as for other, spiritual uses:

> Yes. The exposed roots in the river, when the river washes away some of the gravel, some of the exposed roots will turn bright red ... usually in the water like that, in the current. And you get that ... brew it up into a hair shampoo. Most of the things [like that] you don't boil; just simmer, don't boil it. You boil all of the goodness out of it if you bring it to a rapid boil. You just simmer it ... (January 23, 2011)

Luschiim laughed when asked if the Quw'utsun ever ate the inner bark of cottonwood, suggesting that it would give one bad breath, but then said, "If there was nothing else, yes ... Oh, we have many other edibles; we don't have to eat it, but I do know from other places, it's a food you can eat" (January 23, 2011).

TREMBLING ASPEN (*Populus tremuloides*)

SALICACEAE (WILLOW FAMILY)

HUL'Q'UMI'NUM' NAME: *Qw'iiqw'i'yul'ushulhp* (literally "little dancing tree/plant"; see also lodgepole pine (page 56), which is called *qw'iyul'ushulhp*; Luschiim, January 23, 2011)

DESCRIPTION: Trembling aspen is a slender, white-barked deciduous tree that grows up to 25 metres high and often grows in groups or stands, from the same rootstock. The leaf blades are rounded and finely toothed, and the leafstalks are flattened, allowing them to shake with the breeze. The leaves turn golden yellow in the fall. Male and female flowers are catkins, borne on separate trees. The seeds, like those of cottonwood, are dispersed by the wind.

WHERE TO FIND: Aspens are sporadic but fairly common on Vancouver Island. They tend to grow in moist ground, often with cottonwood and willows.

CULTURAL KNOWLEDGE: Luschiim noted that some athletes bathe in a solution of trembling aspen bark because the tree is so "light," with dancing ability. Also, it doesn't have the same strong odour as cottonwood. Luschiim talked further about aspen:

> It doesn't grow everywhere. So the few places I've seen it growing, it's usually got lots of clay in the ground ... So wherever there's clay in water you're going to find it. So in the past, the ones that used these, that particular tree, you were taught at a young age [where to find it]. For myself, I learned where they are just from walking around, being observant. (January 23, 2011)

He went on to talk about the importance of observation for survival:

> One of the things that's kind of missing, we've lost in this fast world, is we've lost the ability to be observant, [see] everything that's going on. In the past it was, part of it was, a matter of survival. So you watched the weather, the clouds, the wind, the way the birds behaved, how they

sound, what kind of talk they were doing. The way the grass is standing or not standing. You put all these together, you had to be observant, so you know what the weather's going to be like a few days from now. So you know if it's safe to go across the water, four days from now. You plan accordingly, you hunt accordingly, you harvest accordingly. (January 23, 2011)

The shaking of the leaves of the aspen can be one of the indicators for certain weather, "along with a bunch of other things that are happening, or not happening. Including echo. Echo all comes into it. Echo, or lack of it" (Luschiim, January 23, 2011). He also shared that some people use aspen wood for certain spiritual implements, but this is private knowledge.

DOMESTICATED PLUM *(Prunus domestica)*

ROSACEAE (ROSE FAMILY)

HUL'Q'UMI'NUM' NAME:

Plems (borrowed from English)

DESCRIPTION:
Domesticated plum trees are relatively small, bushy and deciduous. Their blossoms are white. The plums themselves vary in size, colour and taste, ranging from yellowish green gage plums to small purple damson plums to sloe plums (*Prunus spinosa*).

WHERE TO FIND: Plums were introduced as a fruit tree from Europe and have been planted around many First Nations settlements as well as on the farms and orchards of settlers throughout the coast.

CULTURAL KNOWLEDGE: Plums were readily adopted as fruit trees by the Quw'utsun. Many people harvested plums and other fruit from abandoned orchards and reserves. Luschiim recalled:

> By 1948, '49, some of the orchards were getting abandoned already. I remember stopping at Cowichan Gap, Porlier Pass—same one, got different names—where the village there was partially abandoned. We went to visit there, and right next to that house was the abandoned ... fruit orchard. There was plums, there was green gages, there were apples, there were pears. We went to visit, and we were, I guess, allowed to pick some. So I don't know, we made a special trip that way, so we could pick on the way. My parents went to visit a man called Plums! [chuckles]. (January 23, 2011)

BITTER CHERRY *(Prunus emarginata)*

ROSACEAE (ROSE FAMILY)

HUL'Q'UMI'NUM' NAME: Bark: ***t'ulum'*** (Luschiim, September 24, 2010, and January 23, 2011); tree: ***t'ulum'ulhp*** (or just ***t'ulum*** or ***t'ulum'***)

DESCRIPTION: Bitter cherry is a deciduous tree, usually relatively small, with distinctive smooth reddish-grey bark that is textured with long horizontal lines. The leaves are small, short stalked and tapered at each end, with fine teeth along the edges. The flowers are small, white and clustered, blooming in spring. The cherries ripen in summer, forming clusters of two to several on a stem. They are small and red when fully ripe. Although they are not poisonous, they taste very bitter, and few people would eat them.

WHERE TO FIND: Bitter cherry grows in open woods and along roadsides and moist edges of meadows from low to medium elevations throughout Vancouver Island and the adjacent mainland.

CULTURAL KNOWLEDGE: The cherries from this tree are small and usually bitter. However, people did eat them as a thirst quencher, as explained by Luschiim:

> We did [eat the little cherries]. Yeah. And it's more stone than berries. Early on they're quite bitter, but if you're up in the mountains and there's no water ... That's one of them [that they used]. The other two are your Oregon-grapes, [and] ... ***shmut'qw'*** ... the orange jelly fungus ... (January 23, 2011)

Bitter cherry bark, ***t'ulum'***, is tough and can be peeled off the tree in horizontal strips. It is used to wrap implements such as bows and to decorate split cedar-root coiled baskets. Sometimes the bark is dyed black to provide a contrasting colour to its natural varnished-red colour. Cherry wood is used to make implements and, because the tree and bark have medicinal qualities, it makes excellent walking sticks that impart health and wellness to the user. As a medicine, for coughs and other ailments, cherry bark can be used alone or mixed with other ingredients:

It's used for different things. It's mixed with a lot of other different things. It can be a wash for external wounds or internal wounds. One frequent mixture for internal [use] would be [with] yew wood, yew tree bark. And **kwulala'ulhp**, your red alder. Many of these medicines, some of them, you've got to be careful with in a way ... like **kwulala'ulhp** will bind you up, if you're using it for some time. So you've either got to stop using it or mix something else in there. So two of the other ones ... are **q'ey'xulhp** (cascara) or willow. Not together though! Causes diarrhoea ... the runs! (Luschiim, January 23, 2011)

Luschiim noted that the quality of cherry bark varies from place to place, depending on the microclimate and other factors. Some families have special places where they go to get their bark. Luschiim said that in a very few places, the bark is so aromatic that you can actually smell it before you harvest it (May 25, 2017). He said there used to be lots of a very special bright-red-barked type of bitter cherry around Westholme: "It got known from all over, so people would come there to harvest them, and then there was no more left." He described how the bark can be harvested sustainably:

So you can harvest just the outer bark, not into the inner bark. Cut lightly, not go into the tree. You go around the stem of the tree, spiral, so you can get a long binding material. If you take it off right, it won't kill the tree. Also, that same material, the bark, was used for decorations on a basket. You could use it [with its] natural colour. You could look for that brighter redder one, or you could dye it to the shade you want, red or darker red or black. (January 23, 2011)

Also, an entire roadside stand of the high-quality bitter cherry was cut down by chipping machines near Westholme. Luschiim described efforts people have made to grow more cherry trees:

I know some people have been trying to grow it at the time and were not successful in getting it to grow from the seeds. That's what they told me. They couldn't grow it. Eventually, I was speaking up ... at Mesachie Lake, at the forestry research station, and they also said they couldn't get it to grow. The reason why we got into that part of the subject was because I was having trouble finding good potent medicine made from

the cherry bark, and I mentioned it to them, and they said they'd been trying to grow some, and they weren't successful. They were guessing that it had to go through a bird before it would germinate. So they were still trying. (January 23, 2011)

He then recalled a time when he learned important information from a man he gave a lift to on how to regenerate bitter cherry by making it grow new shoots—a lesson on how an act of kindness benefits the giver:

So I picked up this hitchhiker, and that's what he does, he works with gardens and growing things ... We got to the subject [of the cut down cherry trees at Westholme] and he said, "Oh, that's easy. You need to get the sprouts, so you can graft to another one." He said, "All you need to do is put a belt around it, a belt, cable chain, about that high [chest high], and the tree thinks it's going to die, and it sends out shoots." So he said you try that, and it'll work. You can get your cuttings off there ... no fruits at the bottom and they're all up there ..." (January 23, 2011)

Luschiim talked about a place upstream from Skutz Falls, on the south side of the river, where there were *t'ulum'ulhp* with particularly bright red ("almost orange") bark, which he was going to go and harvest for walking sticks and various other purposes (January 23, 2011). He noted:

There's very few places where I've gotten really potent *t'ulum'* ... It has a very high aroma, of the same aroma as the others, but has a much stronger smell to it. It has its own definite smell. And when I walked in, all I had to do was open the door and my uncle would say, "Oh, you've got some good medicine!" Within seconds, he says, "Don't forget where you got that one, that's very good ones!" So that one particular spot on Tzouhalem ... there's many spots, but there's one spot in particular ... There's one directly above Skutz, on the south bank, there's one that produces that kind of thing. The other one is up in a place called S-hwuhwa'uselu, Thunderbird Place, up on the south side of the Copper Canyon River ... that's the only place where I came across the one with the high smell, which to Uncle was potent. (January 23, 2011)

GARRY OAK *(Quercus garryana)*

FAGACEAE (BEECH FAMILY)

HUL'Q'UMI'NUM' NAME: *P'hwulhp*

DESCRIPTION: Garry oak is a large deciduous tree, growing up to 25 metres tall, with thick, spreading limbs, and often growing in a gnarled form. The greyish bark has thickly furrowed ridges. The typical lobed oak leaves are shiny and dark green. The pollen-producing male flowers are small catkins, and the female flowers, which eventually develop acorns, are inconspicuous and borne at the tips of the twigs in spring. Some years produce a heavier acorn crop than others. The leaves turn brown and blow off the trees in late fall windstorms.

WHERE TO FIND: Garry oak is almost entirely restricted in Canada to southeastern Vancouver Island and the Gulf Islands. It is common in dry sites and prairies, such as around Somenos Lake, as well as on rocky hillsides and ridges in Quw'utsun territory.

CULTURAL KNOWLEDGE: Luschiim noted that there are two different kinds of Garry oak, just as there are two kinds of camas. One grows in rocky areas and the other in softer ground and prairie lands, putting down deep roots and growing very big:

> Mhm, so it goes with your camas. They [oaks and camas] like the same kind of place. From what I'm told, there's two different kinds. I guess all you've gotta do is look up in Tzouhalem and you see the ones that like to grow on the rocks. And then down here it's the deep roots. They grow in softer ground and they grow big. This is from Siseyutth'e', who was born in 1873 and she died in 1974, 101 years old. That's the date on her grave. Her English name was Agnes Ely Billy. She shared with me that the oak trees were all over Quamichan, hence the camas all over, where it's dry. [There are] some parts that are low and wet. They're not there but they're on shale hills, rocky hills, and that there was many big ones, meaning the oak. [She also talked about] when the sailing ships got here, they come and cut them down. And then they pulled them down to the river with cows. So I would imagine the cows were oxen. But they [the big oak logs] sank. So they cut down big fir trees and they bundled them up together in the summertime; when the river came up in the winter, they were drifted down. Then when they loaded the oaks onto the ships, they just let the fir go. Just drifting around … They took them [oaks] to the old country, yeah [for building ships]. (January 23, 2011)

When asked if he had heard of people eating Garry oak acorns in the old days, Luschiim responded:

> I've heard about it but not from my Elders. My Elders, they talked about somebody else that done it, but they didn't do it themselves. Ah, I know on some of the places, now whether the … deer were digging for, I didn't

check to see what they're digging for. But that's one of the things that, like around now, when you're hunting if it's dry, you listen for **welhts'um'**, dry leaves. The deer, they go pawing, I don't know if they were looking for the acorns or what, but they were looking for something. So you listen and sometimes you luck out that way. (January 23, 2011)

Luschiim pointed out a grey lichen (*Parmelia* sp.) growing on oak trees, which he called **squq-p'iws**: "Flat, wrinkled-up leaves, leafy. **Squq-p'iws**. So here's a good example ... stuck on. So **squq-p'iws** is just stuck on lightly on the bark." This lichen was used as a medicine to treat babies with thrush and was considered an antiseptic (January 23, 2011). (Note: Later, when we went to the Garry Oak Preserve at Somenos Lake on March 9, 2011, Luschiim collected this lichen.)

Luschiim also talked about hunting in oak woodlands:

I liked hunting in the oak tree areas there were, up here on what is now the ecological reserve that is on Mount Tzouhalem. There was always sure to be deer there ... The fit young hunters would hunt up on Tzouhalem, but if they didn't get anything up there, on their way back they would hunt there at the Garry Oak Preserve; usually they would save that area for older hunters who couldn't go up high anymore ... Oaks were a favourite hunting place for me, for several of us, because the deer were sure to be there. And for us, we saved those kind of places for [the end of the day] hunting. If we really needed the deer, we go there. If we don't really need it, we leave it alone. That way we're sure to be able to, if we needed it ,we'd go there ... That was something for us, many of us, we practised ... I guess today we'd call it conservation. So there was certain places like that, or another thing about that. For hunting or harvesting, for those who were able-bodied, meaning in their 20s, 30s, where we had the ability to walk a long ways, we'd go further to hunt. Today I'm 68 years old. If it's the same as it was yesteryear, they'd save those kind of places for us. That's the way it was. (January 23, 2011)

He talked further about the practice of leaving areas for older people to access for harvesting: "I guess you could say it was an unwritten rule, non-written." Today, this principle still applies: "Even today, in modern times—I have a four-by-four vehicle. My brother doesn't have a four-by-four. Even though it's nice hunting in some places, I don't go there, I go further back, so my brother or anyone else who doesn't have a four-by-four could hunt there ... I go further up (January 23, 2011).

Luschiim noted that in some places where the oaks are, the grass sprouts up earlier in the season than in other places, so this was important information for hunters (January 23, 2011).

> A good place [for deer] would be ... Sampson Bluffs: Qwaqwyuqun'
> for description, place Maple Bay and Genoa Bay, on the narrows side,
> there's bluffs there, with moss and grass. They'll be starting to turn
> green now, and **tsqway** is green. So that whole area is a place is for
> hunting; oak trees, lots of oak trees there. (Luschiim, January 23, 2011)

Luschiim also recalled that people used to burn over the Garry oak areas in the winter: "Not too late in the early spring because the camas was already sprouting and would be harmed by the fire, even in January. So they burned in the winter. Also they always watched the weather, and they didn't burn at a high tide time, because at high tide, there was always a pretty strong breeze coming up the slope (say, from the lake), and that would make the fire difficult to control" (January 23, 2011).

People used Garry oak bark as a medicine for wounds (Luschiim, January 23, 2011).

WILLOW (*Salix* spp.)

SALICACEAE (WILLOW FAMILY)

HUL'Q'UMI'NUM' NAME: *Sxwele'ulhp* (applies mainly to Pacific willow, according to Luschiim; January 23, 2011)

DESCRIPTION: There are a number of different kinds of willows native to Vancouver Island, some of which are quite tall, with large trunks, and others more shrub-like. The tree willow species include Pacific willow (*Salix lucida* ssp. *lasiandra*) and Scouler's willow (*S. scouleriana*). Sitka willow (*S. sitchensis*) is also common, as is the fuzzy-leafed Hooker's willow (*S. hookeriana*). The willows are leafy and deciduous, with smooth greyish bark that becomes roughly furrowed with age, and generally narrow or ellipse-shaped leaves. All the different kinds are called by the same Hul'q'umi'num' name, but it seems to apply most prevalently to Pacific willow. Willow flowers are in the form of dense catkins, with the female and male (pollen-producing) flowers on separate trees. The female flowers are wind-pollinated and eventually ripen to release seeds attached to downy parachutes, to be carried away on the wind.

WHERE TO FIND:

Willows are common and widespread in Quw'utsun territory, usually preferring to grow in damp places, such as around the edges of lakes or in marshes, and along creeks or ditches. Some, such as Scouler's willow, grow in open woods.

Pacific willow (*Salix lucida* ssp. *lasiandra*)

CULTURAL KNOWLEDGE: There is a creek running into Kulleet Bay named X̱wála'elhp after "willow tree" (Rozen, 1985, #58). Luschiim said that most people recognize two different types of willow: one that is silvery on the undersides of the leaves and one that is whitish underneath. He noted, "*Sxwele'ulhp* ... the original one is the willow that turns to trees ... turns to trunk. That's the one I know as *sxwele'ulhp*. There's, the way granny put it, granny Celestine, she said the others come from the mountains and kind of slowly make their way down" (January 23, 2011).

Willow branches are flexible. Luschiim described the use of willow bark, or the branches with the bark, to make fish traps and bindings:

> The young ones, of the *sxwele'ulhp*, you use that for making your bindings. And one of our traps for fish is called *sxwu'lu* ... And that word comes from *sxwele'ulhp* and the reason why it's called that is because the bindings are made out of *sxwele'ulhp* ... peel them, comes from the [bark]. It will last for that fishing season. There's no need to harvest something that will last a long time. There's no need to harvest and untie the bindings, so that's what you use ... You could use both [wood and bark]. (January 23, 2011)

Sxwele'ulhp is also used for medicine, but Luschiim cautioned, "You have to be careful not to use too much. Because it's like cascara [laxative]" (January 23, 2011).

Finally, willows can also be used to predict rain. Luschiim noted, "If the leaves turn on edge when there is no wind, it is a sign that rain is coming." This was just one of many different indictors that people would use to help forecast the weather, including bird calls and winds. You have to "put all that together in your head." One clue about weather, for example, is if you see a lot of deer coming out to eat, it may mean that the weather tomorrow will be bad. The deer would know this and would be eating as much as they could to carry them over (Luschiim, September 24, 2010).

Oval-leaved blueberries (*Vaccinium ovalifolium*)

SHRUBS AND VINES

SASKATOONBERRY, or SERVICEBERRY (*Amelanchier alnifolia*)

ROSACEAE (ROSE FAMILY)

HUL'Q'UMI'NUM' NAME: Berries: **tushnets**; bush: **tushnetsulhp** [optional]

DESCRIPTION: A low to very tall bushy deciduous shrub, with greyish bark and bluish-green oval-shaped, short-stalked leaves that turn yellow in the fall. The blades of the leaves are usually toothed around the top half of the margins and smooth around the base. The flowers are white, each with five elongated petals, and borne in clusters of a few to several. They ripen into dark blue or purplish berry-like fruits, which are sweet and edible, usually containing several small brown seeds. Both birds and people enjoy the fruit.

WHERE TO FIND: Saskatoonberry is found on dry bluffs and open woodlands and is common on Vancouver Island and throughout British Columbia.

CULTURAL KNOWLEDGE: Luschiim recognizes four different varieties of Saskatoonberry to be found in Quw'utsun territory, one particularly tall and the others shorter:

> Saskatoon. There's four main kinds here. I only know the one name: **tushnets**. The berries are very sweet. Some of the tall ones that produce the biggest berries, from what I'm told, came from Hwkw'a'luhwum [Qualicum] River. It is a place next to what is now Honeymoon Bay, up at Cowichan Lake. Hwkw'a'luhwum River, today known as Robertson River. And the berries were … That's the name of the river; the name of the place is Ta'mal. And **tatum'**: if I had to warm tea or whatever is warm, it's warm, it's **tatum'**. So that's the name of the place, "warm place." And it's said to produce great big Saskatoonberries. So I went by there several times over my lifetime and yes, they are—unusually bigger than other places. Whether it's because the place is warm, or something special about that place, I don't know. But that's what that place is known for: the larger Saskatoonberries ("huge"). So that's the tall ones.
>
> The shorter ones are from places that are **kw'uluqun**, or shale, a place that has shale … I think it can be almost anywhere. Places on the river, Cowichan River, there's some shale places. And the shrubs never grow more than three or four feet high; some of them are even lower. They're just kind of almost spread out. So that's from one extreme to the other. (January 23, 2011)

Saskatoonberries are well liked, as described by Luschiim, and the Quw'utsun people used to trade for the Interior ones up the Fraser River:

> So the berries were really sought after. There were some [in Quw'utsun territory] but not in great abundance as in the Interior [e.g., Okanagan area]. The Interior ones were sought after because they produced big cakes … we traded clams or something like that for the cakes. Way up, way past Yale, I guess. (January 23, 2011)

Quw'utsun people also dried Saskatoonberries into cakes, but "not as much as the Interior," noted Luschiim (January 23, 2011).

Saskatoon wood is straight and hard, and was used as a material for making implements:

> The wood of the **tushnets** is also very good, similar to ironwood, oceanspray, [in] that you could bend them. The longest ones were used for dipnets, and the shorter ones you could use for hoops or your **p'a'tth'us**, your baby basket. You got your older stems, the older wood. If you're looking for long ones, you went to a place that has a canopy— meaning the trees are like 60, 80, 200 years old, where it's quite a canopy. [When] they're growing under there, they're trying to reach for light, so they grow long, and you can get those ones for hoops, whether it be that [Saskatoon] or whether it be ironwood [*Holodiscus discolor*]. Most use ironwood, but for different spiritual purposes, sometimes they used a different wood. (Luschiim, January 23, 2011)

Saskatoon wood and bark are used for medicine, but it's important to distinguish between it and oceanspray, or ironwood, because they are used medicinally for different purposes. He then described how to tell these two shrubs apart:

> These two look very similar and most people can't tell the difference, but there's one giveaway for this one, and that's there's always ... a U-shape ... There's always going to be a separate change in direction ... roughly a 45-degree angle. If the stem is going like that, there's always a change in direction. There's going to be this angle here, somewhere along the way. And that's one of the ways you can tell. And whether it be a tiny little piece like this or whether it's one like that, you're going to find them. (January 23, 2011)

When asked if he had ever heard of pruning, burning or cutting Saskatoon bushes back (coppicing), Luschiim said, "You can do that to almost anything [with the bushes], and burning is a part of it, yeah" (January 23, 2011).

HAIRY MANZANITA

(*Arctostaphylos columbiana*)
ERICACEAE (HEATHER FAMILY)

HUL'Q'UMI'NUM' NAME: *Qi'qun'aanlhp*, *qi'qunaalhp* ("little arbutus")

DESCRIPTION: A stiffly branching evergreen shrub growing up to three metres tall, with purplish-red bark on the main branches. The leaves are alternate, ellipse- or egg-shaped, two to five centimetres long with short stalks and fine grey hair, especially beneath. The flowers are numerous, small, urn-shaped and clustered. The berries are dark red, small and pithy, not poisonous but not particularly edible.

WHERE TO FIND: Dry rocky outcrops, hillsides and open forests, sporadic in Quw'utsun territory, such as Mount Tzouhalem.

CULTURAL KNOWLEDGE:

This shrub is aptly named as "little arbutus," which it resembles in many ways, and to which it is botanically related. Luschiim commented, "They're kind of scattered around but they're not that common," and that they grow in dry areas. He noted that building roads and ditches creates "microclimates," including locally well-drained areas that are perfect for manzanita. The dry areas on the slopes of Mount Tzouhalem are one place where manzanita grows well, and another is on the southeast end of the Koksilah ridge ... "Way up top is where there's lots." Manzanita has spiritual significance for the Quw'utsun people, although this is not widely known (January 23, 2011).

RED-OSIER DOGWOOD (*Cornus stolonifera*)

CORNACEAE (DOGWOOD FAMILY)

HUL'Q'UMI'NUM' NAME: *Kwum-kwim-mu-mutth'* or *tskwim-kwim-mu-mutth'* when the branches are red (see **stskwim**, "red") or **qw'si'unlhp**, **qw'si'** when the stems are grey or green (Luschiim, January 23, 2011, and November 2017)

DESCRIPTION: A bushy deciduous shrub growing up to six metres tall, with opposite branches that, when young, are often bright red, although sometimes yellowish green. The leaves are oval-shaped, pointed and up to 12 centimetres long, with prominent veins. They turn bright red in the fall. The small white flowers grow in dense flat-topped clusters, and the berry-like drupes are white, sometimes tinged with light blue, each containing a single seed.

WHERE TO FIND: In moist thickets, beside streams and lakes, and in swamps and open forests from lowlands to upper elevations; common throughout Quw'utsun territory and southern British Columbia generally.

CULTURAL KNOWLEDGE: The whitish berries are not considered edible by the Quw'utsun. Luschiim noted that he was discouraged from eating them, but that he had tasted them. He laughed and said, "and they weren't pleasing to me." The shrub often has distinctive red bark. Some people call it "red willow," and a few even call it "red alder," which is the name for the upright tree with red-staining bark. Luschiim commented:

> Some of it's red all the time but more so after a frost, it gets red … It might be worth noting that red-osier dogwood has many other names, used at different communities … It's used for sacred purposes, such as the building of **sh-hwha'hw-thut** sweat lodge [with the flexible branches]. It has [other] spiritual uses. (January 23, 2011)

The leaves and bark are an important medicine, as Luschiim recalled from his childhood:

The leaves and the bark draw out poison. The leaves were used on us when we picked on some bees and they didn't like us disturbing their home. We were probably five, six years old. And we were picking on a beehive, and they attacked us. Three of us got stung really bad. They brought us in, laid us on the floor and Amelia Bob's mom went out and got a whole bunch of those leaves and chewed it and put it on our stings, draw out that poison. So the leaves and the bark you can wash, like a festering wound, with the brewed bark. (January 23, 2011)

The term *p-shut* refers to the action of putting on a chewed-up leaf over a sore or sting as a poultice (Luschiim, January 23, 2011).

HAZELNUT *(Corylus cornuta)*

BETULACEAE (BIRCH FAMILY)

HUL'Q'UMI'NUM' NAME: *P'qw'axw* or *p'qw'áxw* (referring to the nuts or to the whole bush; Luschiim, January 23, 2011)

DESCRIPTION: Hazelnut is a bushy deciduous shrub, up to three metres tall. The leaves are alternate, with blades that are approximately heart-shaped, pointed and doubly saw-toothed at the margin. They turn yellow in the fall. The male flowers are hanging catkins that can be seen hanging from the bushes in spring before the leaves emerge. The female flowers resemble tiny bright red roses. The nuts are hard shelled but delicious, usually produced in twos or threes at the ends of the branches, each surrounded by a bristly sheathing elongated bract.

WHERE TO FIND: Moist open woods; common in parts of southern Vancouver Island and the lower Fraser Valley. Another variety of the same species is common across southern BC. Hazel also occurs along the Skeena River, but its range in the north is restricted.

CULTURAL KNOWLEDGE: Luschiim commented that there used to be more hazelnut bushes around the Cowichan Valley area, and that the squirrels are very fond of them:

> But now they've kind of gone down ... Clarence Elliott, he is probably 10 years older than myself, 10, 12 years. When they used to walk home from school, there was a good tree of this kind. And they had to keep track of the ... squirrels. When the squirrels start to make noise at this particular tree, it was time to go and harvest. So if you didn't go harvest right away, in a couple of days it [the nuts] would be all gone. So they found out they didn't have to keep track of when it was ready to pick. All they had to do was keep track of the squirrels. As soon as they start to make noise there, it was time to go and pick it. (January 23, 2011)

He didn't know why the hazelnuts have decreased. He recalled the old people talking about these bushes and saying they were really big before. He said the flexible branches of hazelnut can be used for making dipnet hoops: "They are not as sturdy or as strong as ironwood [oceanspray], but you can use them" (January 23, 2011). When asked if people used to cut them back to make them grow better, he described a time when they learned how well hazelnut grows after the bushes are cut back:

> I do know they grow *better* when you try to get rid of them. [Laughs.] Yes, some of our youth, not knowing that it was a hazelnut tree, and they went and cleared a whole hillside by the band office, on this side ... And we thought they [the bushes] were gone, but in two years they were all back. And alongside the road, you leave the band office, across the river on that bridge, Indian Road, Miller Road and Allenby Road, you go up Miller Road and the beginning of that hill is all that ***p'qw'axw***. And they cut it down and it grew really good after that. So we don't know if the ones that are growing are the wild kind or the domestic ones [filberts or European hazelnuts] gone wild. (January 23, 2011)

BLACK HAWTHORN *(Crataegus douglasii)*

ROSACEAE (ROSE FAMILY)

HUL'Q'UMI'NUM' NAME: *Metth'unulhp* (Luschiim, January 23, 2011)

DESCRIPTION: Tall deciduous shrub up to seven metres high, with stout, straight thorns usually around 1.5 to 2 centimetres long; bark greyish and smooth on young branches, rough on older trunks. The leaves are alternate, generally oval and three to six centimetres long, the upper end with several jagged lobes, dark green above and quite thick. The flowers are white and five-petalled, relatively small and borne in clusters of several to many. The berry-like fruits are dark, shiny black when ripe, with the remnant calyxes still attached at the upper ends of the fruits; the fruits are edible and sweet but have large seeds.

WHERE TO FIND: Moist areas along streams, lakes and gullies, thickets and open woods from lowland to fairly high elevations. Common on Vancouver

Island, including Quw'utsun territory, and across British Columbia; less common in the north. Extends east to Ontario and south to South Dakota, Wyoming and California.

CULTURAL KNOWLEDGE: The sharp thorns on this shrub, which Luschiim called "pokers," can grow to over two centimetres long. Luschiim commented on the wood: "That's very hard. Makes good tools, such as a fro club, [you have to] dry it, anything for a club like, it should be dried. You can use it green, but it's only going to last a few days (January 23, 2011).

Luschiim also talked about black hawthorn berries, which were not much eaten by Quw'utsun people: "There are other berries, so we didn't really go after these berries [for eating]. But they are a hunger and a thirst quencher. If you are hungry or thirsty, you will eat them because you didn't have anything else to eat. They did have spiritual use, for some people" (January 23, 2011). (Abel Joe had previously mentioned that the charcoal from this shrub is used as a face paint by some people, like that made from devil's-club.)

SALAL *(Gaultheria shallon)*

ERICACEAE (HEATHER FAMILY)

HUL'Q'UMI'NUM' NAME: Berries: ***t'eqe'*** (also applied to whole plant); bush:
t'eqe'ulhp (but not generally used)

DESCRIPTION: Branching shrub, usually one to two metres high, sometimes
creeping. The leaves are evergreen, thick and leathery, elliptical and pointed
and finely toothed around the edges. The leaves are arranged alternately
along the stems. The flowers are small, whitish to pink and urn-shaped,
arranged in a single row along the stalk, usually hanging down. The berries
are purplish black and pea-sized, with finely hairy skin, many small seeds
and dark, staining juice. When fully ripe they are sweet, usually quite juicy
and edible.

WHERE TO FIND: Open Douglas-fir forests to wet rainforest from lowlands to montane areas, forming dense patches; common in Quw'utsun territory and all along the west coast of British Columbia.

CULTURAL KNOWLEDGE: There is a site along the Cowichan River called T'eet'qe' after these berries. Also, Shingle Point IR 4 is called "place of salal berries" (Rozen, 1985, #49). Luschiim explained that although you could say *t'eqe'ulhp* to refer to the bush, "most of the time you would use *t'eqe'*" for the berries or bush. This word is also used to refer to the liver, since the dark colour of liver is the same as the colour of the berries (January 23, 2011). He also discussed the possible origin of the word:

> Some people question where the word *t'eqe'* comes from. Some people say it comes from this word. From here it's in that direction; you know near where the Lhumlhumuluts' longhouse is? Right by the bridge, upstream from that is a place called T'eet'qe'. When logjams formed and rerouted the river and made the river shallow, the T'eet'qe' family didn't come home. They stayed at their camp ... so some say it comes from this word, "camp," T'eet'qe'. (January 23, 2011)

Luschiim mentioned a growth (like **the'thqi'**, the word for a new shoot), such as edible salmonberry or thimbleberry shoots, which can grow on **t'eqe'**:

> So there's a **the'thqi'**-like growth that happens on that **t'eqe'** stem. And that's said to be good eating. I think it's classed as a parasite to that kind. [This is apparently **poque**, or groundcone, *Boschniakia hookeri*.] You can chew on that. (January 23, 2011)

Luschiim had also heard that some people used salal branches in pit-cooking, and that you could chew on the leaves as a thirst quencher.

OCEANSPRAY, or IRONWOOD (*Holodiscus discolor*)

ROSACEAE (ROSE FAMILY)

HUL'Q'UMI'NUM' NAME: *Qethulhp*

DESCRIPTION: A medium to tall deciduous shrub, growing up to four metres tall, with straight, erect branches. The young stems have thin, ridged bark, which becomes greyish when older. The leaves are alternate, broadly triangular and bright green above, paler beneath, strongly veined. The edges of the leaves are shallowly lobed with small secondary teeth. The flowers are tiny and cream-coloured, borne in dense pyramidal, often drooping sprays. The fruits are also small and light brown, the clusters remaining on the bushes over the winter.

WHERE TO FIND: Dry bluffs and rocky slopes to thickets, forest edges and open woods from lowlands to upper elevations; common in Quw'utsun territory and across southern British Columbia.

CULTURAL KNOWLEDGE: Gabriola Passage is called Xwqethulhp, "place of oceanspray" (Rozen, 1985, #37).

The ironwood shrubs growing in the deep forest with long, older stems are what the Quw'utsun use to make dipnet hoops. There are different ages of wood; the younger stems are lighter-coloured and have much more pith than the older stems, even if the older ones aren't much bigger in diameter. Luschiim elaborated:

> The older stems are darker grey in colour ... and so the stems with the darker grey, they come in different shades. The wood you want to use for a dipnet hoop is a darker grey in colour compared to the rest of the others. When you are out there, you look at the bushes, find out the ones that have dark stems, and those are the ones you use for those hoops, about that far apart ... Six feet would be a long one. Long, about that far apart ... Bigger than my thumb, as a base. Maybe the biggest one might be ... two thumbs; that's getting too heavy and too thick ... [So] about the size of your thumb [is best] ... [The dipnet itself is of stinging

nettle fibre.] Yes. Some might make bigger hoops, and some might make smaller hoops.

In conversations about the dipnet, Luschiim described its use and the difficulties of dipnet fishing in both calm and turbulent waters:

For [fishing in] some places it's really hard to use a dipnet, where you lower your dipnet, how you bring it through the water, especially in places that are somewhat calmer, as compared to a turbulent area … [like] Skutz Falls area. That's what I mean by putting it in the water, you pop it into the water. You go down, you bump along the bottom, then you're chasing your fish away. Make sure your edge doesn't hit, that your tail end doesn't drag. When you're fishing in a quiet place, any bump on the bottom can scare your fish away. In turbulent water it's different. At Skutz Falls, I know of two incidents. One boy, a really strong boy, he got yanked into the water with four or five springs [salmon] in his dipnet. And this other boy, he caught several fish in his dipnet and they [the fish] pulled it in; he lost his dipnet. Next time he tied his dipnet to a tree, between his weight and the fish weight in his dipnet, the dipnet broke and he lost his second dipnet. He never came back to that place! (January 23, 2011)

The lesson here is that you don't want to get too many fish in your dipnet; just one or two is all that can be easily handled at one time.

Oceanspray, aptly called ironwood, had multiple important uses for the Quw'utsun: as well as the dipnet hoops, the stems were used for baby cradle hoops, and hoops for the pectin shell rattles of the mask dancers. The stems are also used for salmon barbecuing sticks, *pi'kwun*:

> So I'll tell you a short story on **pi'kwun**. **Pi'kwun** is for barbecuing, one of the kinds to barbecue is your salmon. You split your **pi'kwun**. The diameter of **qethulhp** [is] about 2.5 inches ... put that partway down and leave a long enough end so it'll go into the ground. Then slide your filleted salmon in there, appropriate cross sticks to keep it in there, hold it in there ... lay the salmon in there with the cross sticks. Now, this is the way I was told. If your mom or your granny had to use a dried **pi'kwun**, it'll show other people how lazy you are. "Don't shame me" (that's my great-grandpa talking) "by making your mom or your granny use a dry one." That **pi'kwun** should be moist, green wood. When you see a woman using a dry one, that shows how lazy the man is of that family!

Qethulhp is also an important material for implements, such as digging sticks (**sqelux**), used for clams, camas bulbs or other resources. Luschiim noted, "For digging stick, you burn it to harden it" (January 23, 2011). He noted that the same name is used for the worm-like organ inside a butter clam, as well as the tip or nose of a cockle. He continued, talking about **qethulhp**: "It has a great strength, lengthwise, but the ends do fray if you use it for knitting needles, or, I don't know if it's called frayed, slivers, at the ends, so it wasn't a preferred one for knitting needles ... That **s-tth'ulhp** [mock orange, *Philadelphus lewisii*], that's a preferred one. The other one that's used is **kwunulhp**, which is your ninebark [*Physocarpus capitatus*]" (January 23, 2011). Mock orange is not as plentiful as oceanspray (Luschiim, April 18, 2016). Oceanspray wood was also used for making the long needles (called **tth'qw'e'lhtun** or **putth'tun**) to poke through the reeds when making mats; these needles could be a metre long (April 18, 2016).

COMMON JUNIPER (*Juniperus communis*)
CUPRESSACEAE (CYPRESS FAMILY)

HUL'Q'UMI'NUM' NAME: *Slhelhuq' tsus p-tth'une'yulhp* (literally "lying down flat juniper"; see *slhelhuq*, "lying down," and *stul-'á'tuxw*, "mint"); or the general term for junipers, *p'tth'une'yulhp* (Luschiim, January 23, 2011)

DESCRIPTION: This juniper is a spreading, aromatic evergreen shrub, usually one to two metres tall, with thin, shredding bark. The leaves are needle-like and very sharp, usually around one centimetre long, and spreading around the twigs. The cones are spherical and berry-like, at first green, then ripening to bluish black in the second year. Male pollen cones and seed cones ("berries") are borne on separate shrubs.

WHERE TO FIND: Common juniper grows on dry rocky outcrops and in open woods and coastal peat bogs, from lowlands to alpine areas; in Quw'utsun territory, it is usually found at higher elevations.

CULTURAL KNOWLEDGE: Luschiim said that this low-growing juniper, with its branches lying near the ground, is called generally by the same name as the tall juniper. He said, "Some of those, some are poky and some aren't poky, and I don't know the difference, but some say there's a difference, and they

say, get the poky ones, or get the ones that aren't poky, for the needles" (December 7, 2010). He noted, "Some say you can eat the berries, somebody said they were poisonous; you can eat just a little bit but not too much. One of those, if you're thirsty, anything moist is going to help" (January 23, 2011). Like the seaside juniper, this species also has spiritual uses.

ORANGE HONEYSUCKLE (*Lonicera ciliosa*)

CAPRIFOLIACEAE (HONEYSUCKLE FAMILY)

HUL'Q'UMI'NUM' NAME: *Q'it'a'ulhp*, *q'íte7ey'nulhp* or *q'ut'q'ut'a'ulhp* (see *q'ít'a'*, "swing")

DESCRIPTION: A deciduous woody vine that trails over bushes or climbs in a spiral around the trunks of trees. The leaves are oval or elliptical, usually around five to six centimetres long and opposite, with rounded to pointed tips, smooth or with a fringe of hairs along the edges. The leaves fuse to form a cup beneath the flower clusters. The flowers are bright orange to reddish or yellowish, long and tubular, and clustered several together. Later the flowers ripen into reddish, translucent berries, each with several small seeds.

WHERE TO FIND: Scrambling over bushes or climbing tree trunks in open woods and thickets, from lowlands to montane areas; common in Quw'utsun territory and across southern British Columbia.

CULTURAL KNOWLEDGE: The flowers of this vine are very attractive to hummingbirds. The vines are said to be the swings of ghosts because they seem to move without any visible cause. It is a spiritual plant and has other names that should not be used generally, but only ceremonially. The roots can also be used to make a shampoo. Luschiim noted that "in season you can get male and female roots. In season, meaning when the flower's out, you can tell. Otherwise, you could [use] four roots ... two each of male and female, two and two" (April 16, 2015).

BLACK TWINBERRY, or TWINFLOWER HONEYSUCKLE

(*Lonicera involucrata*)

CAPRIFOLIACEAE (HONEYSUCKLE FAMILY)

HUL'Q'UMI'NUM' NAME: *Shaal'ulhp* (Luschiim, January 23, 2011)

DESCRIPTION: Branching, straggly deciduous shrub growing up to five metres tall, with green, four-angled young twigs and older stems with greyish, shredding bark. The leaves are opposite, with elliptical, pointed blades, usually around 10 to 15 centimetres long, smooth edged and slightly hairy beneath. The distinctive flowers are in pairs, yellow and tubular, cupped by bright red fused bracts. The paired berries are black and shiny, and generally considered inedible.

WHERE TO FIND: Wetlands, lake edges, moist forests and thickets from lowlands to montane areas; common in Quw'utsun territory and throughout southern and central British Columbia.

CULTURAL KNOWLEDGE: Luschiim recalled, "We're told not to bother [with these berries], to leave it alone ... Some say it's edible, but in our part of the country, we were told not to bother [with it]. It may be edible, I don't know ...

It is medicine, but I don't have knowledge of that type of medicine. It's more or less spiritual ... In some places I've seen the stem about this big [five centimetres or more] ... compared to what is around here. That's a big one for around here" (April 16, 2015). The bark and twigs are used medicinally by neighbouring groups, but the berries are universally considered inedible (Luschiim, April 16, 2015).

TALL OREGON-GRAPE *(Mahonia aquifolium)*

BERBERIDACEAE (BARBERRY FAMILY)

HUL'Q'UMI'NUM' NAME: *Luluts'ulhp* (see *luluts*, "yellow")

DESCRIPTION: An upright, few-stalked evergreen shrub, growing to two to three metres or taller, with light-coloured bark, which is bright yellow inside. The leaves are evergreen, alternate and compound with 5 to 11 thick, holly-like leaflets, shiny above, with prominent spiny teeth. The flowers are small and bright yellow in elongated clusters. They ripen into blue, glaucous berries, edible but very tart and with a few large seeds.

WHERE TO FIND: Dry slopes, rocky bluffs, roadsides and open woods from lowlands to higher elevations; common in Quw'utsun territory and across southern British Columbia.

Luschiim explained that the name for this shrub comes from the yellow colour of the flowers. "And if you kind of scratch the skin of the stem or the root, it's also yellow underneath. I've never used it, but I'm told you can make yellow dye out of it ... Almost every part of it is medicine to strengthen the blood" (April 16, 2015).

The berries are edible. Luschiim commented: "I've never bothered to mix them; I could eat them ... they are a thirst quencher when you're out hunting, or whatever you're doing." He also described a game that children like to play: "The berries, before they are fully ripe ... make a game out of it. [When] children are walking they can each grab a handful, similar amounts, and at the signal, they all throw it into their mouths. The last one to make a sour face is the winner!"

Luschiim had heard of the berries being used to treat shellfish poisoning, as was known to some W̱SÁNEĆ people, but didn't know any details, or what they might be mixed with (April 16, 2015).

Both this species and the low Oregon-grape have medicinal, food and spiritual uses, and both were used as a source of yellow dye. The spiritual or sacred uses are considered private knowledge (April 16, 2015).

LOW OREGON-GRAPE, or DULL-LEAVED OREGON-GRAPE (*Mahonia nervosa*)

BERBERIDACEAE (BARBERRY FAMILY)

HUL'Q'UMI'NUM' NAME: Berries: *sunii*, *sunii'*; plant: *sunii'ulhp*

DESCRIPTION: Low, evergreen shrub growing from rhizomes and forming scattered patches. The leaves are compound, each with 9 to 19 leaflets that are evergreen and holly-like, with even, spiny teeth around the edges. These leaves tend to be dull, not shiny like those of tall Oregon-grape, and have a stronger mid-vein. The flowers are small and bright yellow, borne in elongated clusters, and the berries are blue, covered with a waxy whitish surface, pea-sized and edible, but tart with a few large seeds.

WHERE TO FIND: Low Oregon-grape grows in dry to moderately dry sites, on open slopes and in open woods, from lowlands to montane areas and, like tall Oregon-grape, is common in Quw'utsun territory and throughout southwestern British Columbia.

CULTURAL KNOWLEDGE: The berries are edible, although a bit sour. Like those of tall Oregon-grape, they were eaten as a thirst quencher. This plant was used medicinally and spiritually in similar ways to the tall Oregon-grape (April 16, 2015).

INDIAN PLUM, or
BIRD CHERRY (*Oemleria cerasiformis*)

ROSACEAE (ROSE FAMILY)

HUL'Q'UMI'NUM' NAME: *Tth'uxwun'*

DESCRIPTION: A medium, branching, deciduous shrub growing to four metres or taller, with bitter-smelling dark brown bark. The leaves are the first of all the shrubs to emerge in the spring. They are lance-shaped to elliptical or egg-shaped, slightly oblong. The blades are smooth around the edges, light green above and paler below. When crushed, they smell like watermelon rind. The male and female flowers are generally borne on separate plants. Both types of flowers are white, five-petalled and relatively small, arranged in elongated clusters. The flowers emerge very early in the spring, often before the leaves have expanded. The fruits are clusters of small, purple, plum-like edible drupes, each with a large stone.

WHERE TO FIND: Open forests, thickets, roadsides and forest edges at lower elevations; common on southern Vancouver Island, the Gulf Islands and the southern mainland.

CULTURAL KNOWLEDGE: Luschiim noted that this shrub is one of the very first flowers to come out in the spring, but that the berries ripen a little later than some of the other fruits: "The others catch up and pass it, and it's still trying to come out. The salmonberries bloom later and come out earlier than the Indian plum!" The leaves are medicinal. Because they have emetic properties and make one vomit, called **sh-hwuy'utun'um'**, they have been primarily used as a blood cleanser—to help clean out your system. To obtain this effect, a whole bunch of the leaves are chewed. Luschiim remarked that they do not taste good. The plum-like fruits are edible, but as Luschiim said, "it's mostly stone" (April 16, 2015).

DEVIL'S-CLUB (*Oplopanax horridus*)

ARALIACEAE (GINSENG FAMILY)

HUL'Q'UMI'NUM' NAME: *Qwa'pulhp*

DESCRIPTION: This is a distinctive deciduous shrub with stems growing up to three metres tall, about as thick as a broomstick, with numerous needle-like spines. The bark is light brown, the inner bark bright green and the wood soft and whitish-coloured. The leaves are maple-like with seven to nine lobes and spines running along the veins underneath and the leafstalks; they are very large and can measure 30 or more across. The flowers are small, greenish white and arranged in elongated pyramidal clusters, and the berries are small, spiny and bright red, not generally considered edible. The entire plant has a distinctive smell, especially when bruised.

WHERE TO FIND: Moist, rich soil along streambanks and in damp woods, from lowlands to subalpine zone; common in the Cowichan Valley under coniferous forests and widespread across British Columbia.

CULTURAL KNOWLEDGE: Devil's-club is best known for its prickly stems and for its spiritual and medicinal qualities. One or two berries of devil's-club can be eaten as a blood strengthening medicine. The inner bark of the stems and the roots is the usual source of the medicine, which is drunk as a tea, or boiled as a decoction and taken as a tonic for diabetes and various other purposes. Luschiim noted that when you are preparing devil's-club medicine, you generally do not peel the small roots, but if you are using the main stem, you peel it and use the inner bark; the smaller stems are used without peeling, just by scraping off the outer part. Luschiim has noted, however, that recently some people have become very sick from drinking tea made from devil's-club, and cautions against its use (April 5, 2021). In any case, people should always seek the advice of a trained herbal specialist or physician before taking medicines such as this.

FALSE BOX, or
BOXWOOD (*Paxistima myrsinites*)
CELASTRACEAE (STAFF-TREE FAMILY)

HUL'Q'UMI'NUM' NAME: Unknown

DESCRIPTION: A low, erect or spreading evergreen shrub with reddish-brown bark and small, evergreen leaves that are shiny and toothed at the margins. The flowers are small and maroon in colour, clustered along the branches, and the fruits are small capsules, each containing one to two seeds.

WHERE TO FIND: Dry slopes and open woodlands from low elevations to montane areas; common but sporadic in Quw'utsun territory and across southern and central British Columbia.

CULTURAL KNOWLEDGE: Luschiim recalled that when the Catholic priests arrived, they didn't have any palm fronds to celebrate Palm Sunday, so they used false box in place of the palm. But "because they took so many of it, it almost got wiped out" (April 16, 2015). Fortunately, false box is on its way to recovery now. There are lots on some of the Gulf Islands. Florists also buy the branches, "so as soon as they try to come back ... somebody wipes them out ... [but] on some of the islands, the private islands, they're growing nicely because nobody's allowed to pick them" (Luschiim, April 16, 2015).

MOCK-ORANGE (*Philadelphus lewisii*)

HYDRANGEACEAE (HYDRANGEA FAMILY)

HUL'Q'UMI'NUM' NAME: *S-tth'ulhp*

DESCRIPTION: An erect, loosely branching deciduous shrub, growing up to three metres tall, with reddish-brown to greyish opposite, short-stalked leaves. These are three to five centimetres long, pointed and usually toothed at the edges, each with three major veins running from the base. The flowers are relatively large, each with four white petals and bright yellow centres. They are borne in dense clusters and have a sweet fragrance, reminiscent of orange blossoms. The fruits are woody capsules, each about one centimetre long, with numerous seeds.

WHERE TO FIND: Moist woods and thickets to dry rocky slopes, from low to montane elevations. Somewhat sporadic in Quw'utsun territory but occurring on southern Vancouver Island and across southern British Columbia.

CULTURAL KNOWLEDGE: Mock-orange flowers are known for their sweet smell. Mock-orange wood is a preferred material for making knitting needles because, unlike oceanspray (ironwood), it doesn't splinter or fray at the ends (Luschiim, January 23, 2010). Luschiim described its uses:

> So ... *s-tth'ulhp*, the mock-orange ... when you go get that green wood, it'd be workable. It's somewhat hard [to begin with], but within two hours after cutting it down, it starts to harden; even though it's green, it starts to harden. So you only cut enough that you could finish at one time. If you've got a long piece [1 to 1.5 metres long], then if you take enough pieces you can finish, you go lay that piece on the ground. Nowadays we used plastic bags to keep it cool and moist ... Use it for knitting needles, doesn't sliver ... And also used for paddle jackets for dancers [the small wooden paddles hanging from the sleeves and bottom, which rattle with movement], and for tools, like fish net needles [net shuttles], making fish nets, to hold the twine for the net. And there's a couple different ways of making a shape for that. One is like a needle, with two

open ends but the ends are kind of curved in. And the other ... is with a centre ... cut out piece inside. (January 23, 2011)

Luschiim commented that mock-orange wood was one of the preferred woods for the paddle jackets for dancers. A person would not make these of yellow cedar wood unless they were lazy. He said:

So its bark peels off easy, but not as much as ninebark [*kwunulhp*] and not as much as that twinberry, **shaal'ulhp**, not as much as honeysuckle [*q'it'a'ulhp*], but it does peel. The bark is kind of a light brown, kind of a reddish-maroon colour. Sometimes it can be difficult for beginners to tell, but some stems will have a perfect "U" ... and it will have the centre one on there, so it's what I call a perfect "U," and somewhere along the way it'll have a one-sided [branch] but still it's half of a perfect "U." And that's an indicator for ... **s-tth'ulhp**, if you have to get it in the winter, like now [January]. You can't really tell [if] you're not familiar with the different shrubs, then you'll know it. (January 23, 2011)

Mock-orange wood makes excellent barbecue stems. However, Luschiim noted that the shrubs do not live very long, and that it's important to keep track of them, where they are growing, so that you can go and harvest them as needed. A good idea is to take note of them when they are flowering.

Luschiim said, "If you leave it too long, they'll die. They don't like to have a canopy over them. One of the biggest ones I ever saw, about that big [three to four centimetres in width], I waited too long ... by the time I got back to it [after a couple of years], it was too late." You have to "trim it down. Trim it down somewhat, and kind of open some of the canopy ... Something like that yew. If you trim it, it'll last a lot longer" (January 23, 2011).

PACIFIC NINEBARK (*Physocarpus capitatus*)

ROSACEAE (ROSE FAMILY)

HUL'Q'UMI'NUM' NAME: *Kwunulhp*

DESCRIPTION: A medium-sized to tall deciduous shrub with erect branches and distinctively shredding brown bark. The leaves are stalked, the blades with three to five palmate lobes, irregularly double-toothed around the edges. The flowers are small, white and five-petalled, crowded in dense rounded clusters, and the fruits consist of three to five reddish-brown follicles, each containing one to four seeds.

WHERE TO FIND: Ninebark grows in wet to moist soil, along streams and lakes, in marshes and along forest edges, from lowland to montane areas; common in southwestern British Columbia, including in Quw'utsun territory.

CULTURAL KNOWLEDGE: Luschiim said that if you had no other wood available, you might use the wood of ninebark to make needles. He said it makes nice, fancy knitting needles, but the ends can splinter and catch the wool on the needles. The wood is easily carved when still green and is used to make some kinds of tools. It might be used for arrows, but it is too soft and bends or warps easily (April 18, 2016).

LABRADOR-TEA (*Rhododendron groenlandicum*)

ERICACEAE (HEATHER FAMILY)

HUL'Q'UMI'NUM' NAME: Leaves: **me'uhw**; plant: **me'uhwulhp** (Luschiim, November 2017)

DESCRIPTION: An erect to spreading evergreen shrub, growing to one metre or taller, with reddish-brown stems and alternate, leathery leaves. Leaves are elliptical, two to five centimetres long, with the edges rolled under. The undersides are densely fuzzy, white in young leaves, rust-coloured in older leaves. The flowers are small and white, borne in rounded clusters, and the fruits are small ellipsoid capsules.

WHERE TO FIND: Grows in peat bogs, with sphagnum moss, trailing cranberries, sundew and bog laurel, which looks similar but has pink flowers, and the undersides of the leaves are smooth. Not common in Quw'utsun territory on the island, but formerly common on Lulu Island and at Burns Bog. Grows in appropriate habitats throughout BC.

CULTURAL KNOWLEDGE: People make a beverage tea from the leaves. One of the places where these are picked is Ley'qsum, Valdez Island (Luschiim, November 2017).

STINK CURRANT (*Ribes bracteosum*)

GROSSULARIACEAE (GOOSEBERRY FAMILY)

HUL'Q'UMI'NUM' NAME: *Sp'eetth'* (see ***petth'um***, "sour, mousey smell; smell of a tomcat or skunk"; see also seaside juniper, *Juniperus maritima*; Luschiim, April 18, 2016)

DESCRIPTION: A loosely branching deciduous shrub, growing to two metres or taller, with brownish bark and long-stalked, alternate, maple-like leaves, the blades five- to seven-lobed and relatively large—about the size of thimbleberry leaves. The flowers are small and greenish or brownish purple, spaced along a stalk up to 30 centimetres long. The berries are the size of large peas, bluish black with a whitish bloom, and covered with dark spots

(glands). The entire plant has a strong, musky odour, the scent produced from small yellowish glands on the leaves and stems.

WHERE TO FIND: Moist soils along streambanks, lake edges and forested wetlands, from sea level to montane forests; found along the Koksilah and Cowichan Rivers and at higher elevations in Quw'utsun territory, and widespread on the northwest coast of BC.

CULTURAL KNOWLEDGE: This currant is known for its pungent smell. But the berries taste fine if you just put them in your mouth without smelling them. Luschiim described how they are used:

> They can be very juicy ... for a thirst quencher ... I'm told that it grows along the edges of creeks, but if you know where to look, up in the mountains, you can find them up in the mountains, not alongside creeks, but you can find them ... And ... when I'm hunting up there and it's kind of later summer, early fall, you can get them. They are a thirst quencher. That's what we used to look for, a thirst quencher ... There were no drinking bottles before ... the bottles were glass ... we never bothered to carry them, so that was one of our thirst quenchers. Down here in Cowichan, Koksilah area, [these currants are] quite scarce ... and I don't know if the old people of the day wanted to keep it for themselves ... but they told us young ones ... that's a no-no for young people, so ... you got it for the Elders ... (April 16, 2015)

The bark is known as a medicine, but, Luschiim commented, "I wasn't told how to use it ... I was just asked to go and get it" (April 16, 2015).

Luschiim also recognized trailing currant (*Ribes laxiflorum*) and said that some currants had grey-blue berries and some were shiny black. The berries are known to be edible but are sparse (April 18, 2016).

COASTAL BLACK GOOSEBERRY (*Ribes divaricatum*)

GROSSULARIACEAE (GOOSEBERRY FAMILY)

HUL'Q'UMI'NUM' NAME: Berries: *t'em'hw* (Luschiim, April 18, 2016); bush: *t'emhw'ulhp*; branches: *t'em'hwtsus*

DESCRIPTION: A loosely branched deciduous shrub with grey bark, one to three stout spines at the nodes, and sometimes bristly internodes. The leaves are alternate and long-stalked, the blades usually three to four centimetres wide and usually three-lobed, the lower lobes shallowly cleft. The flowers are small with red to white petals and red calyces, growing in drooping clusters of up to four. The berries are smooth, purplish to blackish, up to one centimetre long, with an extended wick from the floral remains.

WHERE TO FIND: Moist thickets, meadow edges and open woodlands and clearings at lower elevations; fairly common on Vancouver Island, including Quw'utsun territory.

CULTURAL KNOWLEDGE: These gooseberries are very flavourful and generally well-liked, although not all bushes have plentiful berries; some have few or none, as noted by Luschiim. He said that the stems are a good medicine, but he didn't know what it was used for. He recalled how children were taught about collecting medicine and assigned the tasks of gathering it as needed:

> As a five-year-old ... that's when my sister Myra and I were shown these plants, and then when the Elders got there, we see them coming. My dad would say, "Okay you go get them." ... He knew what they wanted ... "You go get *t'em'hw*." He showed us these plants closer to the house ... and then when he saw those people coming, I remember him saying, "You remember that plant I showed you?" ... I says, "Yeah." "Well, that's not what you get, you go over there." He described the place by ... by tree or whatever, and you'll find a bush there, and it was our job to go and get it there. And we were five, six, seven [years old]. Children had a lot of responsibility. More than they do today. (April 16, 2015)

GUMMY GOOSEBERRY, or
STICKY GOOSEBERRY (*Ribes lobbii*)

GROSSULARIACEAE (GOOSEBERRY FAMILY)

HUL'Q'UMI'NUM' NAME: *Tum'uqw*

DESCRIPTION: A stiffly branching deciduous shrub with slender spines that are around one centimetre long, brownish or greyish-red bark, and alternate three- to five-lobed leaves. The flowers, single or in pairs, are relatively large, with white or pinkish petals and deep reddish-purple calyces with long, protruding stamens and style. The berries, up to 1.5 centimetres across, are coarsely hairy and glandular, sticky to the touch.

WHERE TO FIND: Open woods, rocky outcrops and thickets from lowlands to upper elevations on southern Vancouver Island and the Gulf Islands.

CULTURAL KNOWLEDGE: The name of this gooseberry is just a little different from that of the coastal black gooseberry. This species has two sets of thorns, slightly sticky leaves and berries that are maroon in colour (Luschiim, April 18, 2016). Luschiim explained the meaning of the word *tum'uqw*: "If you throw a rock in the water from a height, or you throw it up in the air, when it hits the water and it makes a noise: *tu-qwoom'*! That's *tum'uqw* ... That's the word for the sound it makes, *wu tum'uqw*" (April 16, 2015).

Luschiim noted that the berries are "a bit of a chore to eat; it's very pulpy and we peel the skin off it. It's furry like a kiwi [fruit] ... It's maroon in colour. Then you get the pulp, and there's a lot of seeds inside the pulp, so you're spitting out all these seeds. You might get a little bit of pulp ... But we used to eat it" (April 16, 2015).

He noted:

> They are getting very scarce. There used to be lots at the foot of ... what is now called Mt. Tzouhalem, Ts'uw'xilum ... Now there's none. There used to be lots at the reserve at Snaw-Naw-As, Nanoose. Now there's no more there ... I did see a fair amount at ... Kinsol Mines Road, about Mountain Road, over there. So, on your way there [to Kinsol Trestle], maybe two kilometres before you get to Kinsol Mines, I'm guessing now, you'll see Mines Road. So it'll be hanging a right, as you're going in, go on Mines Road, you'll be going under the power line, get past the houses and sometimes the gates are locked, but that's where we used to go on. There used to be a fair amount of them there. (April 16, 2015)

Since gooseberries and currants (*Ribes* spp.) are known to be alternate hosts of the white pine blister rust, it has been recommended in the past to eradicate all gooseberry and currant trees anywhere near white pine trees. This practice may have reduced the numbers of sticky gooseberry from past populations.

RED-FLOWERING CURRANT (*Ribes sanguineum*)

GROSSULARIACEAE (GOOSEBERRY FAMILY)

HUL'Q'UMI'NUM' NAME: Berries and berry-bearing bushes: ***hwihwukw'*** (see
ts'hwikw', "grey"); bush (without berries): ***sqwuliius***

DESCRIPTION: A bushy deciduous shrub growing to two metres or taller, with
erect stems and brownish bark. The leaves are alternate, maple leaf–shaped
but smaller, the blades five-lobed with small teeth around the margins and
fairly long stalks. The flowers are small and bright pink, growing in elongated
clusters that are often drooping. The berries are dark blue with a whitish
waxy coating, covered with small glands. The entire plant has a strong
resinous smell.

WHERE TO FIND: Open woods, along the edges of meadows and roads, and
in thickets; common in Quw'utsun territory and the east coast of Vancouver
Island, from lowlands to montane areas.

CULTURAL KNOWLEDGE: Luschiim
noted that the currant bushes are
called *sqwuliius*, "but when it has
berries the whole thing changes its
name to *hwihwukw'*." The berries
are named after their greyish colour
and are edible. However, the bushes
are treated with caution. Luschiim
commented, "You can eat them …
but … a lady or a girl before child-
bearing or during child-bearing
years is forbidden to go near these …
the whole bush. So we were always
told, don't bring any in the house"

(April 16, 2015). The late Mary Thomas, originally from Westholme and
later from Esquimalt, commented that flowering currant was known as "rain
flowers," and children were cautioned not to pick them, or it would cause rain.

DWARF WILD ROSE, or
BALDHIP ROSE (*Rosa gymnocarpa*)

ROSACEAE (ROSE FAMILY)

HUL'Q'UMI'NUM' NAME: *Xwiinlhp*

DESCRIPTION: A delicate, prickly deciduous shrub with numerous soft prickles and greyish-brown mature stems. The leaves are alternate and pinnately compound, each with five, seven or nine leaflets, and each leaflet around one to two centimetres long, finely toothed around the edges. The flowers are miniature wild rose flowers each with five small pink petals and a five-lobed calyx. The small ovoid hips are orange to red, containing up to 12 "seeds," and lack the calyx segments that remain attached to the hips of the Nootka rose and other wild roses.

WHERE TO FIND: Open woods, forest edges and thickets from low elevations to upland forests; common in Quw'utsun territory and across southern British Columbia.

CULTURAL KNOWLEDGE: This small rose has "several particular uses" in the spiritual realm, making it an important plant; it is ***xe'xe'*** (sacred). The leaves, twigs, petals and hips can be used to make tea, and the outer rind of the hips is edible. You can also eat the ***the'thqi***, or young shoots (Luschiim, April 16, 2015, and November 2017).

NOOTKA WILD ROSE (*Rosa nutkana*)

ROSACEAE (ROSE FAMILY)

HUL'Q'UMI'NUM' NAME: Flowers: *qél'q* or *qel'qulhp*; hips: *qél'eq*; bush: *qel'qulhp*

DESCRIPTION: A medium to tall deciduous shrub, often growing in dense thickets, with stout, erect stems and a pair of large thorns at each node along the stem, sometimes with smaller spines between. The leaves are pinnately compound, with usually five to seven leaflets, each oval or elliptical, measuring five centimetres or longer, coarsely toothed around the edges. The flowers are large and showy, single or two to three together, each with five bright pink petals and a yellow centre. The hips are spherical or slightly pear-shaped and bright orange, filled with "seeds" (achenes) that have irritating slivers around them. At the ends are the remnants of the calyx. The rinds of the hips are edible, and the hips remain on the bushes over the winter.

WHERE TO FIND: Moist thickets, forest edges, hedgerows and open woods from lowlands to upper elevations. Common in Quw'utsun territory, across southern British Columbia and along the coast northwards.

CULTURAL KNOWLEDGE: The name *qel'qulhp* can be used for the flower, but it is used especially if you want to make sure that you are referring to the whole bush. Luschiim recalled, "So as kids we ate the flower ... petals, but also the place where the ... seed [develops; the centre] ... cleaned up good ... That *the'thqi'* [young shoots] is also good, *the'thqi'*, in the springtime ... We peel it ..." He said that one could eat it with the skin on, but he preferred it peeled (April 16, 2015). The orange-coloured rind on the outside of the hip is also edible and is rich in vitamin C (Luschiim, April 16, 2015). You can make tea from the petals, leaves, shoots and hips.

RED RASPBERRY (*Rubus idaeus*)

ROSACEAE (ROSE FAMILY)

HUL'Q'UMI'NUM' NAME: ***Tsulqama'*** (same name as blackcap, *Rubus leucodermis*)

DESCRIPTION: Wild raspberry is a medium-sized deciduous shrub with upright canes that die back after flowering and fruiting in the second year. The stems are prickly, the bark yellowish brown. The leaves are alternate, pinnately compound with three to five sharply pointed leaflets, which are toothed around the edges and prickly along the veins underneath. The flowers are white, borne in clusters of two to four, and the berries are bright red, with multiple drupelets forming the typical raspberry, one to two centimetres wide, which falls intact from the white receptacle when ripe.

WHERE TO FIND: Moist thickets, clearings, rocky slopes and forest edges. Not well known on Vancouver Island, but found throughout BC, mostly east of the Coast and Cascade Mountains.

CULTURAL KNOWLEDGE: Wild red raspberry is not found on Vancouver Island today, but Luschiim talked about a botanist who was in Quw'utsun territory around 1840 and reported seeing lots of red raspberries (***tsulqama'***) growing in the area, and a Russian botanist who wrote a report in 1860, just 20 years later, who said there were very few. (Luschiim recalled that this same botanist also said there used to be a lot more deltoid balsamroot in the region than there is now, or was even in 1860 when he was there.) Luschiim knows that there are plenty of wild raspberries up in the Interior, around Merritt, for example, because he has gone hunting up there and seen them. Luschiim speculated that the Quw'utsun people were trying to plant raspberries on their lands around 1840, but the berries didn't take well and had more or less disappeared within a few decades. The berries are edible and delicious (Luschiim, April 16, 2015).

BLACKCAP, or
BLACK RASPBERRY (*Rubus leucodermis*)
ROSACEAE (ROSE FAMILY)

HUL'Q'UMI'NUM' NAME: Berries: ***tsulqama'*** ("raspberry"; also used for red raspberries)

DESCRIPTION: An erect, deciduous shrub with very prickly arching stems, the bark light bluish purple. The leaves are alternate, compound, with three to five pointed leaflets, saw-toothed around the edges, bright green above and paler beneath. The flowers are white and five-petalled, in clusters of two to seven, each with many stamens. The berries are raspberry-like but dark bluish purple, usually sweet and juicy.

WHERE TO FIND: Rocky slopes, clearings, forest edges and thickets, from lowland areas to higher elevations; common in Quwut'sun territory and in southern British Columbia.

CULTURAL KNOWLEDGE: The berries are edible and very good. Luschiim commented, "You learn where to pick them, meaning they're nice and sweet in a sunny area, but in a shaded area they're not as delicious" (April 16, 2015). They are easy to dry for winter use, alone or mixed with other berries. Luschiim said they were usually just dried singly, not mashed together (April 16, 2015). A tea can be made from the tips of the leaves, and/or the roots of blackcap. Usually you need to break off the leaf tip because it's dried up (Luschiim, September 24, 2010).

THIMBLEBERRY, or
REDCAP *(Rubus parviflorus)*

ROSACEAE (ROSE FAMILY)

HUL'Q'UMI'NUM' NAME: Berries: ***t'uqwum'***; edible shoots: ***lhequ*** or ***the'thqi'***
(also applied to salmonberry shoots); plants: ***t'uqwum'-úlhp***

DESCRIPTION: A medium-sized shrub, usually under two metres, often
forming patches. The stems are erect, without spines or prickles, and with
shredding brownish bark. The new shoots are bright green and mildly
fragrant. The leaves are large and maple-like, with long stems and five to
seven lobes, saw-toothed along the margins and covered with soft hairs
on both sides. The flowers are relatively large and showy, each with five
white petals, and growing in clusters of 2 to 10. The berries are fine-grained,
raspberry-like, bright red and readily dropping from the receptacle when ripe.
They are sweet and juicy when growing in a moist area.

WHERE TO FIND: Moist open areas, thickets, roadsides and open woods from
lower elevations to montane areas; common in Quw'utsun territory and
throughout southern British Columbia.

Thimbleberry leaves and flowers

CULTURAL KNOWLEDGE:
The young sprouts, or
shoots, of thimbleberry
(*lhequ*) can be snapped
off in the spring,
peeled and eaten raw.
Also, when these are
at the right stage for
eating (around the
middle of May), it is
an indicator that it is
the best time to cut
cedar poles (about
15 centimetres at the
butt end). These are

peeled and split to create the foundation for coiled cedar-root basketry. Luschiim elaborated:

> You'd be looking for a very specific time [for getting these poles]. One of those is the **the'thqi'** of the redcap? When that **the'thqi'** is ready to eat, that's when you harvest cedar pole, roughly six or eight inches [across]. You can use the morning side of that pole to make the fine boards of your cedar basket [the foundation]. Well, it could be as thin as the thickness of this pen, pencil, maybe a little wide for some, and you rip that into your mini-lumber. That's your strength for your basket, inside of it. So, like the bones of a basket. So when that **the'thqi'** is ready to eat, that's when you harvest that. You can get it any time of the year, but you're going to lose a lot of it. So there's several things that have indicator plants … [This would be] … probably May, and it'll vary, even a little bit from here to Cowichan Bay, there's a difference. And Saanich would be … quite a bit earlier down there, yeah [maybe around April]. And some of them, you harvest cedar bark at the time of the full moon, yeah. Some families are that way. (December 16, 2010, with Trevor Lantz; also May 25, 2017).

Ripe (red) and ripening (pink) thimbleberries

SALMONBERRY (*Rubus spectabilis*)

ROSACEAE (ROSE FAMILY)

HUL'Q'UMI'NUM' NAME: Berries: ***lila'*** (also applied to the bush); edible shoots: ***the'thqi'***

DESCRIPTION: Medium to tall shrub, growing to three metres or higher, with stiff branches that have strong, sharp spines, and shredding, light brown bark. The leaves are raspberry-like, each with three pointed leaflets, saw-toothed

Salmonberry leaves and flowers

around the edges. The flowers bloom early in spring, before the leaves expand. They are showy, each with five bright pink petals, and produce sweet nectar that attracts hummingbirds and bees. The berries vary in colour from golden yellow to deep ruby. They resemble large raspberries, 1.5 to 2 centimetres long, falling off the receptacles when ripe. They are one of the earliest of the fruits to ripen, from late May to July.

WHERE TO FIND: Moist soils in swamps, thickets, road edges and open forests at low to high elevations; common in coastal British Columbia.

CULTURAL KNOWLEDGE: The young sprouts can be broken off, peeled and eaten in the springtime, like those of thimbleberry. The berries come in a variety of colours, depending on the genetic makeup of the bushes, from yellowish to salmon-coloured, bright red and dark red. They are a favourite fruit of the Quw'utsun and others who live in coastal BC, and one of the earliest berries to ripen.

Salmonberries of different colour forms

TRAILING WILD BLACKBERRY
(*Rubus ursinus*);
HIMALAYAN BLACKBERRY
(*R. armeniacus*);
and EVERGREEN or
CUTLEAF BLACKBERRY (*R. laciniatus*)
ROSACEAE (ROSE FAMILY)

HUL'Q'UMI'NUM' NAME: *Sqw'iil'muhw* (applied to any kind of blackberries, including the trailing blackberry and the larger, introduced types); Himalayan blackberry: *xwum sqw'iil'muhw* or *xwum'xwum' sqw'iil'muhw* ("ripens early" blackberry); evergreen, or cutleaf blackberry: *'ayum sqw'iil'muhw* ("ripens late" blackberry)

DESCRIPTION: Trailing blackberry is a prickly, woody vine growing to four metres or longer, with thin stems that sprawl over bushes or trail along the ground, rooting at the tip. The bark is smooth and light bluish or greenish. The leaves are compound, raspberry-like and 10 to 15 centimetres long with three to five segments, each pointed and coarsely toothed around the edges. The leafstalks and midveins on the undersides of the leaves are also prickly. The flowers, produced in small clusters, are five-petalled and white. Male and female flowers are on separate plants, with only the female ones producing berries. The berries themselves are black, sweet and juicy, ripening usually in July, well before the introduced Himalayan and cutleaf blackberries. Himalayan blackberry is much more robust, ripening later in July and August, with larger, more rounded berries, and forming dense, impenetrable thickets. Evergreen or cutleaf blackberry ripens even later. Its leaves are darker green and with deeply cut lobes.

WHERE TO FIND: Trailing blackberry is common in open woods, hedgerows, thickets and clearings, usually at lower elevations; it is widespread on Vancouver Island, including in Quw'utsun territory. Himalayan and evergreen blackberries grow in disturbed open ground, roadsides, old fields and thickets.

Trailing wild blackberry (*Rubus ursinus*)

CULTURAL KNOWLEDGE: Trailing blackberry, the native type, has sweeter, earlier-ripening fruits, which are well liked and widely harvested by the Quw'utsun and others who live within its range. Blackberry leaves are used to make a tea, which is also a kind of medicine (Luschiim, April 16, 2015). Luschiim mentioned a story about a jealous husband who killed his wife, and from her blood, the trailing blackberry originated.

BLUE ELDERBERRY (*Sambucus cerulea*)

CAPRIFOLIACEAE (HONEYSUCKLE FAMILY)

HUL'Q'UMI'NUM' NAME: *Tth'uykwikw*

DESCRIPTION: A very large bushy shrub or small tree, often over five metres tall, with a thick trunk and pithy twigs. The leaves are pinnately compound, divided into five to nine elliptical, pointed leaflets, each around 10 centimetres or longer and with saw-toothed margins. The flowers are small and cream-coloured, borne in dense, flat-topped clusters, and the berries are small and powder-blue in colour, covered with a heavy waxy coating, each containing three to five seeds. The berries are juicy and tangy in flavour. This species blooms later than the red elderberry, and the berries ripen in late summer and early fall.

NOTE: The leaves, stems, bark and roots of elderberries are poisonous, and if they are used medicinally, should only be prepared under the guidance of an herbal specialist. Only a few berries at a time should be eaten raw; normally it is best to cook them.

WHERE TO FIND: Open areas along roadsides and in thickets; sporadic in Quw'utsun territory and on eastern Vancouver Island and the Gulf Islands, but common in south central British Columbia.

CULTURAL KNOWLEDGE: Luschiim mentioned that this shrub "likes its feet or toes dry," whereas the red elderberry is more tolerant of dampness (April 16, 2015). Luschiim explained the name is related to that of "blue jay" (Steller's jay), which is *skwitth'uts*, named after the light blue colour, *kwikwutth'*. The lighter-coloured jays have a name, *skwi-tth'uts-alus*, similar to that of blue elderberry (*tth'uykwikw*).

Both the blue elderberry and the red elderberry are considered to be good blood strengtheners—almost all parts of these bushes. Luschiim said the berries of both are edible, but he had never heard of drying them, just eating them fresh, or usually cooking them. The stems of both can be used for making toys. The pith in the centre of the stems is very thick and can be easily removed: "you can work them out of there, get rid of them, and then you've got a pea-shooter ... so that's what they were used for" (Luschiim, April 16, 2015).

Luschiim has observed that animals, including deer, will only eat certain individual plants:

> It's interesting. Many of the plants, including these two [elderberries] ... you can travel up the mountains and you can see whole valleys of these ones, whatever they may be, plants or individual plants, or a few individual plants—the whole valley can be covered with them, but the deer or elk or other animals will only eat a certain one or certain ones. My question is, is it to their taste or is it medicinal uses? The reason why I say that is because in the past, many years ago when I used to bring medicine to Elders, the old people, certain elements—needles or bark. As soon as I walk in, they were making comments like, "Oh, that's a good medicine, son, don't forget where you got it!" In the sense it's the same way as the [animals] ... certain ones are good. Over time I've been keeping track of where I got those [medicines] that all seem to be a pattern of several valleys come together; that's where the good ones were. I don't have any scientific backing for something like that, but there it is. (April 16, 2015)

RED ELDERBERRY *(Sambucus racemosa)*
CAPRIFOLIACEAE (HONEYSUCKLE FAMILY)

HUL'Q'UMI'NUM' NAME: *Tth'iwuq'*

DESCRIPTION: Large bushy deciduous shrub, growing to five metres or taller, with thick, soft, pithy twigs and dark reddish-brown bark on the mature stems. The leaves are opposite and pinnately compound, with five to seven pointed elliptical and lance-shaped leaflets, finely toothed at the edges. The flowers are small, cream-coloured and arranged in pyramidal clusters. The berries are small and bright red (occasionally yellow or dark-coloured) and each contains small seeds.

NOTE: The seeds of red elderberry can release cyanide when chewed, which can cause nausea and vomiting, so the berries should only be eaten after cooking. The leaves, stems, bark and roots are also poisonous, and if used medicinally, should only be prepared under the guidance of an herbal specialist.

WHERE TO FIND: Moist open woods, forest edges, roadsides and thickets from lowlands to upper elevations; common in Quw'utsun territory, on Vancouver Island and across southern British Columbia.

CULTURAL KNOWLEDGE: Some people consider these berries to be poisonous; they should not be eaten raw, but formerly people cooked them into a sauce and ate them. Some people also used them medicinally, but they should only be prepared by someone with expert knowledge, as mentioned above. If the stems are used as pea-shooters, they should be dried, not used fresh.

SOAPBERRY, or SOOPOLLALIE

(*Shepherdia canadensis*)

ELAEAGNACEAE (OLEASTER FAMILY)

HUL'Q'UMI'NUM' NAME: Berries: *sxwesum*; bush: *sxwesumulhp*

DESCRIPTION: A medium-sized deciduous shrub with erect to spreading branches with brownish, scaly bark. The leaves are opposite, smooth-edged and elliptical, up to six centimetres long, dark green on the upper surface and lighter green with brownish scaly spots beneath. The flowers are small, with a four-lobed greenish-yellow calyx and no petals. The flowers are clustered at the ends of the branches and appear before the leaves in spring. Male (pollen-bearing) and female (berry-bearing) flowers are borne on separate plants. The berries, bright red or reddish-orange, oval, and six to eight millimetres long, are juicy, bitter and soapy to touch when crushed.

WHERE TO FIND: Medium to dry sites from sea level to montane regions; occurs on southeastern Vancouver Island; sparse but common on limestone sites, at edges of woods and clearings.

CULTURAL KNOWLEDGE: The berries are well known for their foaming properties, due to a natural detergent. The Quw'utsun and other First Peoples of British Columbia whip the berries with water into a frothy dessert, still enjoyed today. Usually sugar or another sweetener is added. The berries cannot come in contact with any oil or grease, however, or they will not whip.

Luschiim said that **sxwesum** was formerly common, until the cement plant at Bamberton was built: "Used to be lots but when ... the cement plant went in at Malahat, it killed them off" (April 16, 2015). They were common on the Malahat, but also farther toward Duncan:

> There should be a whole bunch [soapberry] ... right next to the reserve and maybe in the reserve ... at Malahat. I know, next to the highway between the highway and Malahat reserve, they were going to turn it into a limestone pit. They were going to mine limestone there. I don't know what happened. It fell through ... So you're going up the Malahat, before you get to the top, there's a powerline on your right ... goes up

the hill to a house or whatever up there. Right at that spot, there'll be a cement barrier and ... as soon as that runs out, there's a place to park and there's a log sticking out of the rock bank; you'll see some bushes there ... Right by that ... what we call a gun barrel log sticking out ... That's where you'll see it. There's a few bushes around there ... (April 16, 2015)

Luschiim noted that **sxwesum** is also a medicine for the Quw'utsun (April 16, 2015).

HARDHACK, or SPIRAEA (*Spiraea douglasii*)

ROSACEAE (ROSE FAMILY)

HUL'Q'UMI'NUM' NAME: ***T'eets'ulhp***, "fish spreader plant" (see ***t'eets'***, "fish spreader sticks" or "cross-sticks"; see also snowberry, *Symphoricarpos albus*; Luschiim, April 16, 2015)

DESCRIPTION: A low to medium deciduous shrub with erect stems growing from creeping rhizomes and often forming dense thickets. The slender twigs are usually somewhat woolly. The leaves are alternate, short-stalked and oval, with coarsely toothed edges; they are dark green and smooth above, paler and sometimes hairy beneath. The flowers are small and pale to deep pink, borne in attractive, dense pyramidal clusters. The fruits are dry, small and brownish follicles.

WHERE TO FIND: Moist, open ground in swamps, bogs, ditches and lake margins from lowlands to montane areas; common throughout Quw'utsun territory and all of British Columbia.

CULTURAL KNOWLEDGE: The branches are hard and somewhat springy, making them a suitable material for making fish spreaders "because you have to be able to bend them a little to insert them, but then they have to spring back and hold the fish flat ... Flexible, possibly the one side a little bit bent, they work better for me" (Luschiim, September 24, 2010). Luschiim noted that there are two kinds of spiraea: "One that's more pointed, a little bit wider at the bottom and more pointed at the top, that's more common. The other one, here ... these are around but not as common as the other one [with the pointy flowers] (April 16, 2015)." A site on the Cowichan River is called Sht'eets'elu, meaning "place to get fish spreaders [from hardhack]."

WAXBERRY, or
SNOWBERRY *(Symphoricarpos albus)*
CAPRIFOLIACEAE (HONEYSUCKLE FAMILY)

HUL'Q'UMI'NUM' NAME: Berries: ***p'up'q'iyaas*** (can also be used for the whole bush); bush: ***p'up'q'iyaasulhp***; straight stem suitable for fish spreader: ***t'ets'ulhp*** (see also hardhack, page 173; Luschiim, September 24, 2010)

DESCRIPTION: A bushy medium-sized deciduous shrub, often growing in dense patches, with greyish-brown bark and opposite elliptical to oval leaves, sometimes with a few wavy teeth or irregular lobes. The flowers are small, pink and urn-shaped, growing in short dense clusters, and the berries are waxy-white, soft and clustered, often persisting on the twigs over the winter. The berries are not edible and have a reputation for being toxic.

WHERE TO FIND: Waxberry grows in thickets in open woods, the edges of meadows and roadsides, and disturbed areas from lowlands to montane zones. It is very common across southern British Columbia and throughout Quw'utsun territory.

CULTURAL KNOWLEDGE: The straight branches of waxberry are called *t'etsu'ulhp*, after *t'eets'*, the word for fish spreader sticks or cross-sticks (Luschiim, September 24, 2010).

Luschiim noted that the sticks from this shrub can be used for *t'eets'* (fish spreaders), since they are just the right size and the right flexibility. They can also be used as skewers for smoking clams after they have been steamed in a cooking pit, but the "proper" skewers would be made from *qethulhp*, oceanspray (April 16, 2015). Luschiim remarked, "Sometimes, you know, maybe you are kind of in a rush and you don't have the time to go and whittle down a good stick [of oceanspray] ... we'd use that" (April 16, 2015). He also said that they use the berry and the stems for a game, flipping the berries with the sticks: "It's all part of coordination, eye-hand movements, body movements, coordination. You've got a berry, get a stem from there, and you learn which kinds of stems are nice and springy. Some of them don't spring very well ..." Waxberry also has many spiritual uses for Quw'utsun people (April 16, 2015).

A similar species, creeping waxberry (*Symphoricarpos occidentalis*), is also known to Luschiim. He said it has a similar Hul'q'umi'num' name to those of pearly everlasting (*Anaphalis margaritacea*) and kinnikinnick (*Arctostaphylos uva-ursi*)—something like *tl'ikw'un'* (April 16, 2015).

HUCKLEBERRIES and BLUEBERRIES (*Vaccinium* spp.)

ERICACEAE (HEATHER FAMILY)

There are several different kinds of blueberries and huckleberries occurring in Quw'utsun territory. Sometimes the same names are applied to slightly different species. The main types are listed here.

Red huckleberries (*Vaccinium parvifolium*)

CANADA BLUEBERRY, or VELVET-LEAVED BLUEBERRY (*Vaccinium myrtilloides*) and BOG BLUEBERRY (*Vaccinium uliginosum*)

ERICACEAE (HEATHER FAMILY)

HUL'Q'UMI'NUM' NAME: *Maal'sum'* (identified as "cranberry" in some dictionaries, but likely Canada blueberry and/or bog blueberry)

DESCRIPTION: A low deciduous shrub up to 40 centimetres tall, growing in dense patches. The branches of Canada blueberry, especially when young, are covered with dense, velvety hairs. The leaves are alternate and elliptical or lance-shaped, up to four centimetres long, with smooth edges and pointed tips. Bog blueberry leaves are bluish green and more oval or oblong. The

Bog blueberry (*Vaccinium uliginosum*)

flowers bloom when the leaves are half-grown. They are whitish or pinkish, bell-shaped and in terminal clusters (Canada blueberry) or borne singly (bog blueberry). The berries of both are edible and sweet, up to one centimetre across, and blue with a pale whitish bloom.

WHERE TO FIND: Dry forests and clearings in sandy or rocky soils, or peat bogs; uncommon in southwestern British Columbia, except in the bogs of the Lower Mainland and Fraser River delta.

CULTURAL KNOWLEDGE: Luschiim said that people used to pick the berries called **maal'sum'** in the Fraser Valley, in the Fraser River delta around the Quw'utsun village of Tl'uqtinus (April 16, 2015). This term could refer to either Canada blueberry (which has clustered berries) or bog blueberry (berries borne singly), both of which grow in peat bogs in the Lower Mainland and have sweet and juicy berries. Bog blueberry also grows in Vancouver Island bogs.

Bog blueberries and Labrador-tea

ALASKA BLUEBERRY (*Vaccinium alaskaense*) and OVAL-LEAVED BLUEBERRY (*Vaccinium ovalifolium*)

ERICACEAE (HEATHER FAMILY)

HUL'Q'UMI'NUM' NAME: *Lhew'qím'*

DESCRIPTION: Erect to spreading deciduous shrubs, up to two metres tall. The young twigs of Alaska blueberry are usually yellow green; those of oval-leaved blueberry are usually reddish to reddish-brown. The leaves are alternate; Alaska blueberry leaves are finely serrated, oval and pointed; oval-leaved blueberry leaves are blunt at both ends and usually smooth around the edges.

The flowers of both species are pinkish and urn-shaped, borne singly in the leaf axils. Oval-leaved blueberry flowers usually appear in spring before the leaves have fully expanded; Alaska blueberry flowers appear when or

Alaska blueberry (*Vaccinium alaskaense*)

after the leaves expand. The berries of oval-leaved blueberry are pleasantly sweet and juicy, pea-sized, sometimes globe- or pear-shaped, and dark blue or purple with a distinctive whitish waxy bloom, giving them a light blue appearance. Those of Alaska blueberry are shiny, dark purple or blackish and can be somewhat tart.

WHERE TO FIND: Both Alaska and oval-leaved blueberries grow in open forests, clearings and the edges of creeks and wetlands from lowlands to subalpine regions; common across the coast of BC and throughout Quw'utsun territory.

CULTURAL KNOWLEDGE: Luschiim had heard about a kind of blueberry called **lhew'qim'** from an interview with Abner Thorne. Abner wasn't sure which type it was, but thought it was the one whose berries were "really close to the colour of a harbour mussel or bay mussel shell, **lhew'qum'**." This would seem to be oval-leaved blueberry, whose berries are indeed a dusty blue, due to a waxy coating on the skin (Luschiim, April 16, 2015).

Oval-leaved blueberry (*Vaccinium ovalifolium*), see also photo on pages 120–121

BLACK MOUNTAIN HUCKLEBERRY (*Vaccinium membranaceum*)

ERICACEAE (HEATHER FAMILY)

HUL'Q'UMI'NUM' NAME: *Yey'xum'* (the name for "a mountain blueberry"; see also evergreen huckleberry, page 183)

DESCRIPTION: A short to medium-sized, erect deciduous shrub, densely branched and growing up to 1.5 metres high. The young twigs are yellowish green, and the leaves are alternate and elliptical, growing to five centimetres or longer, and rounded to tapered at the base, narrowing to a pointed tip. The margins are finely and evenly toothed. The leaves turn bright red in the fall. The flowers are pale yellowish pink, solitary and urn-shaped, borne in the leaf axils. The berries are up to one centimetre across, globe-shaped and purple, with or without a pale bloom. They are sweet and edible—one of the choice berries of this group.

WHERE TO FIND: Dry to moist coniferous forests and clearings, usually at upper elevations; common throughout British Columbia, and occurring in Quw'utsun territory at higher elevations.

CULTURAL KNOWLEDGE: Luschiim said that the berry called *yey'xum'* grows up in the hills, such as at the top of Mount Prevost (Swuq'us), as well as on Mount Arrowsmith (April 16, 2015). It is widely known as a sweet and flavourful berry, highly appreciated. A related species, Cascade bilberry (*Vaccinium deliciosum*), also has sweet berries and may also be called *yey'xum'*. (See also evergreen huckleberry, page 183.)

EVERGREEN HUCKLEBERRY

(*Vaccinium ovatum*)

ERICACEAE (HEATHER FAMILY)

HUL'Q'UMI'NUM' NAME: *Yey'xum'*
(see also black mountain
huckleberry, page 182)

DESCRIPTION: An upright to
spreading evergreen shrub, up to
three metres tall, with alternate,
leathery, shiny leaves that are
oval-shaped with pointed tips
and sharply toothed margins.
The flowers are small pinkish,
narrowly bell-shaped and borne
in small clusters. The berries are
small, spherical and found in two
genetically distinct colour forms:
one deep purplish black and

shiny and the other light blue due to a waxy whitish coating. They ripen in fall
and can often be picked into early winter.

WHERE TO FIND: Open dry to moist forests from lowlands to montane areas;
common in certain places on southern Vancouver Island, including in parts of
Quw'utsun territory, especially on limestone.

CULTURAL KNOWLEDGE: People apparently used to go out to Sooke, the
territory of the T'Sou-ke Nation, to pick these berries, although Luschiim
was not really familiar with them (April 16, 2015). Luschiim is uncertain as
to whether the name *yey'xum'*, which was used by W̱SÁNEĆ Elders for this
huckleberry, might pertain instead, or as well, to black mountain huckleberry
(see page 182). Luschiim noted that there were lots of evergreen
huckleberries at Jordan River, and that people harvest the green branches
to sell to florists as greens in flower arrangements, especially from around
October to December (April 16, 2015).

BOG CRANBERRY (*Vaccinium oxycoccos*)

ERICACEAE (HEATHER FAMILY)

HUL'Q'UMI'NUM' NAME: ***Qwum'tsal's*** (see ***qwum'st***, "so sour it makes you close your eyes"; ***qwum'st tun'qulum*** means "close your eyes"; Luschiim, January 23, 2011, and November 2017)

DESCRIPTION: A tiny shrub with slender, creeping stems and very small evergreen, oval, alternate leaves, deep green and shiny on the upper surface and greyish beneath with margins rolled under. The flowers, single or in small clusters, are small and pink, with four reflexed petals and prominent anthers and stamens. The berries are spherical or ovoid, one centimetre or longer and bright red, nestled into sphagnum moss. Ripening in the fall, they are edible and quite tart.

WHERE TO FIND: The vines are nestled in the moist sphagnum moss of peat bogs and muskeg areas throughout BC, including Quw'utsun territory such as, formerly, around Tl'uqtinus on Lulu Island in the Fraser River delta.

CULTURAL KNOWLEDGE: There is a place called Qwumtsulasum, "place of cranberries," at the northwest corner of Reserve No. 1, past the hospital, on the Cowichan River. These berries are also found at Qul'i'lum', Dougan Lake (formerly Dougan's Lake) at Cobble Hill: "There used to be lots of *qwum'tsal's* from what old people [were] talking about" (Luschiim, April 16, 2015). They were formerly found in the meadows at landowner Ed Nixon's property at the north end of Langford Lake, in the area called "Turner Meadows." The berries also grow in an area on the mainland as one heads toward the US border, as described by Luschiim:

> But there was many places we went to ... Now, let's head toward the US border. You come off at Tsawwassen, you get to the main highway and now you're on your way to the US border. You're on a flat, the bay is on your right ... And pretty soon you're going to start to climb the hill. Before you climb the hill, all that area was a place for getting *qwum'tsal's*, according to the old people ... some call it Campbell River; I don't know why ... You know my family harvested there on my dad's side, probably on my mum's side too. But my dad's side comes from ... [near] Blaine ... Then from there they moved to Semiahmoo, on a long spit near Blaine. So at the beginning of that spit on the shore side, that's Xwul-xwuluqw. That's where my great-grandmother's [Luxlaxulwut's] family came from. (April 16, 2015)

The berries are tart but were, and still are, enjoyed after cooking.

RED HUCKLEBERRY (*Vaccinium parvifolium*)

ERICACEAE (HEATHER FAMILY)

HUL'Q'UMI'NUM' NAME: *Sqw'uqwtsus* (used for both berries and plant)

DESCRIPTION: An erect deciduous shrub growing up to four metres high, with bright green, angled branches. The leaves are alternate, oval and two to three centimetres long, with smooth edges. Leaves on the very young plants are evergreen and have finely toothed edges. The flowers are whitish to pink, urn-shaped and borne singly in the leaf axils. The berries are globular, up to one centimetre wide and bright red. They are edible and juicy but somewhat tart.

WHERE TO FIND: Open woods, often growing from stumps or rotten logs, from lowlands to montane zones; common in western British Columbia, including throughout Quw'utsun territory.

CULTURAL KNOWLEDGE: Luschiim described the derivation of the name, *sqw'uqwtsus*:

> So one of the ways to harvest is to hit the berries off branches ... That's why it's called **sqw'uqwtsus**. You put some kind of material underneath to catch it. There's other ways, one by one, or you can use like a, a type of comb, rake type of thing ... Comb it off ... In September in some places there'd be lots ... Up in the mountains ... and maybe even August, mid August, but they have maggots or worms inside. So for us, for my family, we only eat the early ones ... We don't bother with the later ones. (April 16, 2015)

People used to dry these berries in cakes, sometimes mixed with other kind of berries. Nowadays they are used for jam and jelly, pies and muffins.

HIGHBUSH CRANBERRY *(Viburnum edule)*
CAPRIFOLIACEAE (HONEYSUCKLE FAMILY)

HUL'Q'UMI'NUM' NAME: *Qwemtsuls* (Luschiim, May 2017)

DESCRIPTION: A sprawling to erect deciduous shrub spreads from rhizomes and from natural layering and rooting of the lower branches. The bark is smooth and greyish to reddish. The leaves are opposite and elliptical, often with three coarse lobes or jagged teeth. The flowers are small and white, borne in rounded to flat clusters, and the fruits are pea-sized, bright red when ripe, and clustered, each with a single large, flattened seed.

WHERE TO FIND: This shrub grows in moist areas along streambanks, in swamps and open woods, from lowlands to montane areas; found throughout BC, but uncommon in Hul'q'umi'num' territory, especially at lower elevations.

CULTURAL KNOWLEDGE: Luschiim hasn't seen many highbush cranberry bushes around. He noted that he had seen one highbush cranberry bush growing halfway up Mount Tzouhalem (Squw'utsun; January 23, 2011, and May 25, 2017). He noted that it seems like you can almost see through the berries if you hold them up to the light. He didn't know of a name for them and hasn't used them very much (April 16, 2015).

Great camas (*Camassia leichtlinii*)

HERBACEOUS
FLOWERING PLANTS

YARROW *(Achillea millefolium)*

ASTERACEAE (ASTER FAMILY)

HUL'Q'UMI'NUM' NAME: *T'uliqw'ulhp*

DESCRIPTION: An aromatic herbaceous perennial with erect, slender stems up to one metre tall but usually shorter. The leaves are finely divided and fernlike, the lower ones stalked, the upper ones unstalked. The flowers heads are numerous, forming a flattened or rounded cluster. The ray flowers are white or sometimes pinkish-tinged, and the tiny disk flowers are cream-coloured. The fruits are single seeds, dry achenes.

WHERE TO FIND: Medium to dry clearings, including meadows, coastal bluffs, rocky slopes and edges of woods or roads. Common and widespread throughout BC, including in Quw'utsun territory.

CULTURAL KNOWLEDGE: Yarrow is an important medicinal plant, used to treat colds and other ailments. Luschiim harvests yarrow year-round because he can recognize the leaves and stalks even in the winter. He would go to places like Tzouhalem to get this medicine. "If someone lets me know, I can go ... even in the dark and I can get it right now ... can be back before the dance is finished and I'll have a bag for you. So that kind of knowledge of knowing where to get it is very important" (June 24, 2011).

WILD NODDING ONION (*Allium cernuum*); TAPERTIP ONION or HOOKER'S ONION (*A. acuminatum*); and GARDEN ONION (*A. cepa*)

LILIACEAE (LILY FAMILY)

HUL'Q'UMI'NUM' NAME: *Qw'exwíyuts*, *qw'exwiyuts* (possibly tapertip onion, or both types; Luschiim, January 23, 2011). **LUMMI NAME: *Sqwulniich*** (wild nodding onion; Al Scott Johnny, April 6, 2021)

DESCRIPTION: Onions are known by their distinctive odour. They grow from bulbs and have thin, grass-like leaves. Nodding onion bulbs are pinkish-skinned and elongated, and those of tapertip onion are small and globular. Garden onion bulbs are much larger and usually spherical or ovoid-shaped. The bulbs are covered by a thin, papery skin. Nodding onion flowers are small and pink, in a rounded cluster borne on a stalk that bends downwards: hence the name. Tapertip onion flowers are small and reddish purple, borne in a rounded, upright cluster. Nodding onion plants often grow in clumps. Onion fruits are small, three-lobed capsules, and the seeds are black in nodding and tapertip onions.

WHERE TO FIND: Nodding and tapertip onions grow in dry to medium-dry clearings, rocky bluffs, upper beaches, grassy slopes and open forests from sea level to montane elevations. Both are found on southeastern Vancouver Island, including Quw'utsun territory.

CULTURAL KNOWLEDGE: Luschiim was told by his great-grandfather Luschiim that their people used to eat wild onions (January 23, 2011). Some Quw'utsun people he spoke to did not think that this was accurate, but later Luschiim found that his

Wild nodding onion (*Allium cernuum*)

great-grandfather's information was confirmed in 19th-century explorer Robert Brown's journals (September 24, 2010). He explained:

> Oh yeah. I knew from Luschiim that we did use it [onions]. Ah, again, when the modern onion got around, you know, they're much bigger and easier, so the use of it kind of fell to the wayside. But when I started asking Elders around here, they said, "Oh no, we didn't use it." So I was kind of at a loss until I read Brown's journal. When he went up the [Cowichan] river, Loxe's son, Loxe Jr., went along as a guide and a cook. Because I was reading Brown's journal, in that book it says that man came down from hunting, and he had eggs, grouse eggs, with him, and he had wild onions. And they cooked, I think it was a stew, with those eggs and the wild onions, so that supports what Luschiim told me. (January 23, 2011)

Luschiim said he would ask Roy Edwards, an Elder who has wide cultural knowledge (along with Augie Sylvester and Luschiim), about the name **qw'exwiyuts** and to which plant it is applied. He was adamant that the name **speenhw**, which some have identified as nodding onion, is applied only to edible blue camas (January 23, 2011).

Tapertip onion, or Hooker's onion (*Allium acuminatum*)

SPREADING DOGBANE, or
HEMP DOGBANE (*Apocynum androsaemifolium*)
APOCYNACEAE (HEMP DOGBANE FAMILY)

HUL'Q'UMI'NUM' NAME: *Qelutsulhp* (*qeluts'*—used for any spinning material)

DESCRIPTION: Spreading dogbane is an herbaceous perennial that spreads from rhizomes and forms patches. The stems are often reddish, and the plants grow to half a metre or taller, with spreading branches. The leaves and stems exude a milky juice, or latex, when cut. The leaves are opposite along the stem, spreading and drooping, oval or elliptical in shape and around five to eight centimetres long. They are smooth above and usually hairy on the undersides. The flowers are small, pink and urn-shaped, in flat-topped clusters, and the fruits are long cylindrical pods that split open to reveal numerous parachuted seeds.

WHERE TO FIND: Clearings, roadsides and open woods in well-drained, dry, sandy soil, from lowlands to montane areas; common throughout BC, including in Quw'utsun territory.

CULTURAL KNOWLEDGE: Liz Hammond-Kaarremaa, who is a spinner and weaver, talked to Luschiim about this plant, which he said was formerly used for spinning. He said that the stems are harvested just before the leaves start turning yellow (mid-August). It has very tough fibres in its stem. Luschiim told her that it grows at the gate to the Duncan rifle range (where she was able to locate it), and by the washrooms at the entrance to the Kinsol Trestle along the Cowichan Valley Trail (August 11, 2016).

According to Captain Charles Wilson (1866), the tall Indian hemp (*Apocynum cannabinum*) was traded from upriver as "Fraser river hemp" and was used for fish nets, duck nets and fishing lines, as well as blanket warp. It would have been known to the Quw'utsun from the village of Tl'uqtinus on Lulu Island. However, Luschiim was not aware of a taller kind of "hemp," although he confirmed that many items and materials were traded up and down the Fraser River by the Quw'utsun and others.

RED COLUMBINE (*Aquilegia formosa*)
RANUNCULACEAE (BUTTERCUP FAMILY)

HUL'Q'UMI'NUM' NAME: Name not known to Luschiim

DESCRIPTION: A herbaceous perennial growing from a taproot. The leaves are finely divided, with multiple rounded lobes; the basal leaves have long stalks and the stem leaves are fewer and shorter-stalked. The flowers are attractive, in loose clusters, each with five distinct red-spurred petals, yellow sepals and long stamens. The fruits are five erect, pointed follicles and the seeds are small and black.

WHERE TO FIND: Moist meadows, clearings, roadsides and open woods throughout BC, including throughout Quw'utsun territory.

CULTURAL KNOWLEDGE: Luschiim said this is "a very important spiritual plant; very important to our people." He added, "I looked for many decades for the name for that, because there are spiritual uses, and maybe that's why that name wasn't shared, because it was higher value and spiritual use. Only those people that ... specialize in spiritual ... Maybe were allowed to know that name ..." Luschiim said there were at least five names for this plant, but that they were mostly not used except for spiritual purposes. He said, "So up in the mountains, certain mountains have a lot of them ... One is Arrowsmith. Lots of them ... I've seen lots ... at Mount Todd and Mount Modeste, and close to Port Renfrew, and all along the creek there at Gordon River camp, where it's shady" (April 16, 2015).

WILD GINGER (*Asarum caudatum*)

ARISTOLOCHIACEAE (BIRTHWORT FAMILY)

HUL'Q'UMI'NUM' NAME: ***Tth'uletth'le'*** or ***tthuletth'le'een'*** ("heart leaves"; Luschiim, June 24, 2011)

DESCRIPTION: A creeping herbaceous perennial, growing from rhizomes and forming extensive mats. The dark green leaves are stalked, with heart-shaped to kidney-shaped blades that are finely hairy with prominent veins. The flowers, usually hidden under the leaves, are purplish brown and bell-like, with three spreading, pointed lobes. The fruits are flesh capsules with small ovoid seeds.

WHERE TO FIND: Moist, shaded forests from sea level to montane zone. Widespread in moist areas across southern BC; occurring in a few locations in Quw'utsun territory.

CULTURAL KNOWLEDGE: This plant is a very special one for the Quw'utsun people. Luschiim noted, "The stem is sweet tasting and could be chewed on for a kind of candy. Also, it has important spiritual uses. There are only certain places where it grows—in sort of shady spots" (April 16, 2016). Luschiim said that this plant is becoming quite rare locally, and he is careful not to share the locations where it grows, because some people might go and take all of it. He noted that sometimes people mistake wild lily-of-the-valley (*Maianthemum dilatatum*) for wild ginger, since both plants have heart-shaped leaves. When asked if people might have transplanted wild ginger from one place to another to make it more accessible, he commented,

> "We transplanted pretty well everything there is, everything that grows, plant or animal; from early on, I was told we planted salmon eggs from one stream to another ... Berries, yes. Seafood, clams, oysters, yes. We moved them ... Whatever was somewhat rare then, you'd transplant them; certain important plants were moved, such as ... wild ginger." (April 18, 2016)

DELTOID BALSAMROOT

(Balsamorhiza deltoidea)

ASTERACEAE (ASTER FAMILY)

HUL'Q'UMI'NUM' NAME: Not recalled by Luschiim

DESCRIPTION: red listed: This is a "red-listed" species, very confined in its distribution. It is an herbaceous perennial growing from a deep taproot with a woody stem base. The basal leaves are long stalked, with triangular "deltoid" blades that are green above and lightly whiteish and hairy beneath, with rounded teeth. The stem leaves are smaller and less numerous. The flowerheads are sunflower-like, with bright yellow rays and yellow disk flowers. The outer bracts on the heads are slightly woolly. The fruits are small, smooth achenes, resembling miniature sunflower seeds.

Deltoid balsamroot in a meadow with blue camas, buttercup and clover flowers. The deltoid balsamroot flowers are large and dark yellow.

Deltoid balsamroot blooms

WHERE TO FIND: Dry grassy meadows, bluffs and open woodlands, often in rocky areas; rare on Vancouver Island—found only in a few locations, including Mount Tzouhalem (Squu-'utsun) in Quw'utsun territory.

CULTURAL KNOWLEDGE: Luschiim said that the taproots of this "sunflower" were formerly pit-cooked and eaten, but they haven't been used for some time, and not many people know about their former use. Luschiim said that there was a Russian botanist in the area who wrote a report in 1860, noting that there used to be lots of "that small sunflower" earlier, but that even by 1860 there were very few left (April 16, 2015). He also said that deltoid balsamroot grows in very particular sites:

> Camas grows in a pretty dry area, so there's not many [other plants] that grow there. That deltoid balsamroot, where I've seen them, they grow in kind of mini ravines; they don't grow out where the **speenhw** [camas] is. There's a dip there ... that's where you're going to get your sunflower. They need, from what I've seen, they need a certain amount of shade, whereas **speenhw** can grow out in the open. That's one of the places we need to take a walk at. There's about three patches at Mount Tzouhalem ... and outside, there's one bigger patch and two smaller patches outside of the fenced area. There must be more by now." (January 23, 2011)

CALYPSO, or FALSE LADYSLIPPER (*Calypso bulbosa*)

ORCHIDACEAE (ORCHID FAMILY)

HUL'Q'UMI'NUM' NAME: *Ti'tuqw-el'tun*, *ti'tuqw-el'tun'* (may pertain to "growing straight up"; Luschiim, June 24, 2011)

DESCRIPTION: A perennial herb growing from a white, fleshy corm, with upright stems usually 10 to 15 centimetres tall. The leaves are single, positioned at the base, elliptical to oval in shape, with parallel veins. The flowers, which bloom in early summer, are single and showy, with bright reddish-purple sepals and petals and a lower slipper-shaped pouched petal, whitish and spotted or streaked with darker purple. The fruits are ellipsoid capsules, producing numerous tiny brown seeds.

WHERE TO FIND: Mossy forest floor from lowlands to upper elevations, forming small patches; occurs across BC, but quite rare in Quw'utsun territory.

CULTURAL KNOWLEDGE: Luschiim suggested that the name possibly pertains to the restricted opening part of the flower. He noted that it is an important spiritual flower, and people used the bulbs as well. He said that it has become very rare since the Cowichan Valley was logged, but now it is starting to come back as some of the second-growth forests are maturing. There are certain places where it is common (Luschiim, June 24, 2011).

COMMON CAMAS (*Camassia quamash*)
and GREAT CAMAS (*C. leichtlinii*)

LILIACEAE (LILY FAMILY)

HUL'Q'UMI'NUM' NAME: *Speenhw* (see page 192; Luschiim, January 23, 2011)

DESCRIPTION: These two species are similar, both growing from egg-shaped tapering bulbs with bright green, upright, grass-like leaves and blue flowers arranged in a raceme along an upright stalk. The flowers are large, each with six tepals (petals and petal-like sepals), six stamens, and a single three-parted pistil. Common camas blooms about two weeks earlier than great camas, around mid-April, and its stalks are shorter—usually around 50 centimetres tall, as compared to great camas at 80 to 100 centimetres tall. The tepals of common camas are slightly asymmetrical, with five spreading outwards and upwards and one pointing down. Those of great camas are arranged symmetrically. When the tepals wither, those of common camas

Common camas (*Camassia quamash*)

remain spreading, whereas those of great camas twist up around the developing seed capsule. The three-parted fruiting capsules are barrel-shaped in common camas, more three-sided in giant camas. Both produce brown, shiny seeds.

WHERE TO FIND: Both species grow in moist open meadows and clearings, mostly at lower elevations, and are common on southeastern Vancouver Island and in Quw'utsun territory. They were maintained by periodic burning and also managed by selective harvesting and by timing the harvest to allow the plants to fruit and produce seeds.

CULTURAL KNOWLEDGE: The month of May is called Punhwe'num and is "the time when the blue camas blooms" ("Hul'q'umi'num': About Our Language," 2000, p. 114).

Luschiim said that people recognized the two different kinds of camas—a taller one and a shorter one—and that they generally preferred the larger one (great camas), "but you can also eat the smaller ones" (September 24, 2010). He hasn't harvested camas very much himself but noted that if you saw any death camas (page 257), you would pull it out to avoid confusing the bulbs with those of edible camas:

Wherever they come up, you take them out. Make sure they didn't get a hold. I see them up on Tzouhalem. They're coming more and more ... Camas grows in a pretty dry area, so there's not many [other plants] that grow there. That

Great camas (*Camassia leichtlinii*)

deltoid balsamroot, where I've seen them, they grow in kind of mini ravines; they don't grow out where the **speenhw** is. (January 23, 2011)

He also said that if there was no camas in a place, or if only the smaller kind grows there, you might transplant bigger ones there instead.

He noted:

A good place [for camas] would be ... Sampson Bluffs, for lack of a name for them: Qwaqw-yuqun' ... Maple Bay and Genoa Bay, on the narrows side, there's bluffs there. They'll be starting to turn green now [January] and **tsqway** is "green." So being bluffs, with moss and grass, it'd be starting to turn green now, and **tsqwaqwiyaqn** is a green place. So that whole area is a place is for hunting; oak trees, lots of oak trees there.

Luschiim described how people used to burn over the ground to promote the growth of camas and certain berries:

In the Cowichan Valley, there's some at Quamichan ... And Quamichan stretched to the stone church, all the way to what is now the Somenos

Camas bulbs (left) and wapato tubers

Lake. And there and other places, we burned the ground area every few years. It comes from many different Elders, including great-grandfather Luschiim. So ... along with *speenhw* in those kind of places, there's also other berries, such as *sqw'iil'muhw*, trailing blackberries, and your blackcaps and strawberries. So what [Great-Grandpa] Luschiim shared with me: after a few years, the ground would turn "sour"—that's what he called it, *ni' sa'yum' thut tu tumuhw* ... So you burn the ground, burn the grass and other vegetation to sweeten the ground—that's his words, "sweeten the ground"—so that the ashes will fertilize the ground. He said that the strawberries would get really tiny, but after you burn, the berries will be the size of your thumbnail. So for camas they went far, again from Luschiim, he said, we have places way past *stth'a'mus*, Beacon Hill, and we had several islands out in the Gulf Islands that were exclusive to different families; these islands were exclusive to certain families. (January 23, 2011)

Luschiim continued to describe camas and its use:

You could keep *speenhw* in a basket. Preparation for a special feast: [in] some families, only the men cooked the *speenhw*. Away from the village. And one of my grandfathers, Sulxulh—his name was Sandy Jones—showed me a place where, out a certain point, where there was hundreds of years they'd been cooking *speenhw*. Where the rock was indented from burning; it's got an indentation there. Certain years there was not as much *speenhw*, I was told, and [at] those times, the *speenhw* we reserved at feasts, for certain dignitaries that we would invite over. It was also used for trade, was what I was told, with the mainland. Some of the trade was sometimes not in actual changing hands, but some places we brought *speenhw*, or clams, or dried fish, we brought them to our families over there in the Fraser River. They watched over our garden, like what they call the *sqewth* [wapato, *Sagittaria latifolia*], so they'd guard it, but when it was close to harvest time we'd be there, and we'd bring along other foods that they want or need. Having said that, though, some of those places over there, they also had a camp here in Cowichan. They could dig camas here too. (January 23, 2011)

Great camas blooms

Luschiim said you could cook the bulbs and dry them, or just dry them fresh: "It could be either way ... You might bring some. But camas ... some of them were cooked along with, for example, the **s'axwa'** [butter clams], (steam) pit cooked **s'axwa'**, you layered them between the camas; **s'axwa'** is your clams, could be **s'axwa'**, or it could be a mixture of all the different clams. Flavoured." (January 23, 2011)

When asked when people stopped eating camas so much, Luschiim responded,

That really would be a guess on my part. But you can imagine the amount of work to gather camas and other vegetables we use, compared to the new vegetables that came. Well, look at the size of a camas compared to a potato. Some of them could be six inches long. Well, you can imagine a camas compared to a potato, 3.5 inches in diameter. The new vegetables slowly came in and we slowly stopped eating the others, such as the camas ... the other one being the deltoid balsamroot, the sunflower [*Balsamorhiza deltoidea*]. I understand there used to be lots here in Cowichan, in the valley, but by 1860s they started disappearing according to a botanist [possibly Robert Brown; see page 192] … So for camas, I guess maybe we can use that time of 1860s [when the potatoes started to come in], yeah. (January 23, 2011)

When asked if he could see camas being revitalized as a food in the future, he responded, "I think the modern term would be 'exotic.' I can see it coming back as an exotic food, as a special food. I know some people are trying to bring it back. And I wouldn't mind. I should make the time to try to start a garden of camas and others …" (January 23, 2011). He said it would be a good start if the Nature Conservancy's Garry Oak Preserve at Somenos were open to Quw'utsun people coming to harvest camas there, but noted that this should be accompanied by tending and burning:

One of the things is … any plants … when a piece of land is neglected … the ground gets packed, and when the ground is packed, it won't grow. So you need to start turning it over, and once it's turned over, it will grow … If it could be worked out somehow to do some burning, not little

Common camas and western buttercup (*Ranunculus occidentalis*)

plots that are 3 × 3 [feet]: plots 30 × 30 [feet]—that would be a small plot. Do several of them in one year and keep going from there ... You have to know what you're doing. It's something you grew up with it. If you didn't grow up with it, you have to learn how to do it ... We did burning at Quamichan all the time. No fuel, no fire. So I lived at a place where it's a hill. When it's dry, it's dry. So we'd burn the grass. But you gotta know when to burn it ...

The worst time to burn would be about one o'clock in the afternoon. Everything's dry by then, [it would burn] five acres, that's how hot it is. So if you burn early in the morning ... Or, you can do the edges at different times, maybe when the first time it's dry, burn the edges [so] you're going to have a perimeter ... so you're going to have a stopper. Keep track of the weather: how hot it's going to be. You can definitely do some in, maybe, in a shower. So you can burn it in a little shower, you can coax it along. Not in the summertime. Your fuel's going to be gone in by summertime. Any time now [January], winter, yeah winter; by the time spring's come around, you're too late. You're going to stunt the growth. So, springtime, you're much too late. This is January. But we have had some warm weather already, so I would say it's too late, because the tops [of the camas] will be coming up now ... It'll be about two weeks ahead of here [in Victoria]. Still, this is mid-January, [it's] too late ... [to burn]. (January 23, 2011)

Luschiim had pit-cooked camas bulbs only once, and there were only four bulbs. He noted, "You can add things to it to make a nice colour. They change colour ... pink, if you add something, but I don't know what that is ..." (Luschiim, January 23, 2011; the Straits Salish people add alder bark and arbutus bark to colour the camas bulbs when cooking.)

When asked if the camas patches were owned by particular families, Luschiim responded, "That'd be more of a guess. But like any other thing, there'd be certain patches that would be owned by a family or an individual from that family, and others that're owned by a community, or several families of a community" (January 23, 2011).

Some kinds of animals, probably deer, like to eat camas leaves. Luschiim has noticed that the tops are gone in some places (January 23, 2011).

TALL BASKET SEDGE *(Carex obnupta)*

CYPERACEAE (SEDGE FAMILY)

HUL'Q'UMI'NUM' NAME: *Tl'utl'* (possibly also **p-shey'** for some; see page 225)

DESCRIPTION: A patch-forming perennial herb with stems growing to one metre or higher and dark green, tough, upright leaves, two to five blades per stem, with very sharp edges, somewhat channelled at the midrib and pointed at the tip. The flowers are dark brown spikes in clusters of three to seven, the terminal one with many male (pollen-bearing) flowers, the lower spikes with mostly female flowers. The fruits (perigynia) are egg-shaped and leathery, containing small, smooth, lens-shaped achenes.

WHERE TO FIND: Shaded forest swamps, sloughs and wet meadows, forming extensive patches. Common on BC's southern coast, including in Quw'utsun territory.

CULTURAL KNOWLEDGE: The leaves of this sedge are split in halves and used for weaving baskets as the twining elements. Some people use it for baskets in Quw'utsun territory, although the Nuu-chah-nulth weavers are the main users of this material. This sedge also has several spiritual uses (Luschiim, June 24, 2011).

PIPSISSEWA, or
PRINCE'S PINE (*Chimaphila umbellata*)

ERICACEAE (HEATHER FAMILY)

HUL'Q'UMI'NUM' NAME: *Ququn'alhp*

DESCRIPTION: A slightly woody perennial herb, with stems around 15 to 20 centimetres tall, growing from rhizomes and forming small patches. The leaves are evergreen, leathery, elongated and arranged in whorls of three to five, each leaf up to six or seven centimetres long and sharply toothed with pointed tips and short stalks. The flowers are pink and nodding, borne in a terminal cluster, and the fruits are globular capsules, each containing numerous seeds.

WHERE TO FIND: Moderately dry open forests in middle elevations; found occasionally in Quw'utsun territory in mossy areas.

CULTURAL KNOWLEDGE: Using the suffix *–alhp* as an indicator, Luschiim suggested that this plant is medicinal (April 16, 2015). Some people use the plant as a general tea, for everyday use. Some people say you shouldn't use too much of it, but Luschiim doesn't know about this.

Luschiim noted that the introduced weedy shrub called daphne laurel (*Daphne laureola*) looks a lot like this plant when it is young, and you have to be careful not to confuse them, since daphne laurel is poisonous. He said, "I've seen them [daphne laurel] at the Juan de Fuca gymnasium in Colwood. There's fields behind the building that's just full of that [plant] and some of them are a person high. I can't stay there too long because of the toxicity in the air" (June 24, 2011).

SHORT-STYLED THISTLE (*Cirsium brevistylum*)
and other thistle species (*Cirsium* spp.)
ASTERACEAE (ASTER FAMILY)

HUL'Q'UMI'NUM' NAME: ***Xuw'xuw'iinlhp*** (general name for thistles in the area; originally applied to native thistles; Luschiim, June 24, 2011)

DESCRIPTION: Short-styled thistle is a short-lived perennial herb growing from a taproot. Its stems are erect, hairy and simple or somewhat branching above. The basal and lower stem leaves are up to 30 centimetres long and broadly lance-shaped, wider above the middle; the upper stem leaves are shorter. The leaves are lobed to nearly smooth-edged, with slender yellow spines along the margins; they are hairy, especially above. The

Short-styled thistle (*Cirsium brevistylum*)

flower heads are borne in clusters of three to six at the top of the stalks; the bracts surrounding the heads are densely cobwebby, slender and tapering. The multiple disk flowers crowded into each head are purplish red, sometimes lighter, and the fruits are achenes, each attached to a light-coloured "parachute." Another native thistle is edible thistle, *C. edule*. Common introduced thistles include Canada thistle (*C. arvense*) and Scottish thistle (*C. vulgare*). All are similar, although the flowerheads of Canada thistle are more numerous and smaller than the others.

WHERE TO FIND: Wavy-leaved thistle grows in moist meadows, open woods and roadsides from lowland to upland sites; frequent on the southern coast of BC, including Quw'utsun territory.

CULTURAL KNOWLEDGE: Luschiim pointed out that the name, *xuw'xuw'iinlhp*, pertains to the growth pattern of the thistles in January or February; their basal leaves are spread out from the centre like a starfish:

> So it's a starfish with many legs; it [the sun star] usually grows pretty big … in January or February the native thistle is like that … on the ground, spread out [like a multi-armed starfish]. And right now, at about 1,000-foot level, that original thistle is about this high [45 centimetres or so] and you can tell the difference between the native ones and introduced ones. (June 24, 2011)

He said eating the young leaves of the thistle in the spring helped strengthen the immune system. He also noted that thistle had other spiritual uses: "That's one of the top medicines. And being out in January and February, way before other plants are out, makes it a top medicine." He recognized the name used by W̱SÁNEĆ, Lummi and Nooksack people, *s'eyeth-ilhch*, which is derived from the word *s'iiya*, meaning "sharp," and noted that some Quw'utsun people might use that word for thistle if they had W̱SÁNEĆ or Lummi ancestry (June 24, 2011).

PACIFIC HEMLOCK-PARSLEY (*Conioselinum gmelinii*)

APIACEAE (CELERY FAMILY)

HUL'Q'UMI'NUM' NAME: *Shewuq* (also the name for garden carrot, *Daucus carota*)

DESCRIPTION: A smooth perennial herb growing from a short taproot or cluster of fleshy roots. The stems are solitary and can grow over one metre tall. The leaves are finely divided and fern-like. The flowers are small and white, in multiple umbrella-shaped, rounded clusters. The fruits are oval, five to eight centimetres long, ribbed and with broad, thin wings. The entire plant has a carrot-like scent.

NOTE: This plant is very similar to poison hemlock (*Conium maculatum*), a toxic, weedy plant in the same family. There are other poisonous species with a similar appearance, including water parsley (*Oenanthe sarmentosa*) and water hemlock (*Cicuta douglasii*); anyone wishing to use Pacific hemlock-parsley should take great care not to confuse it with these poisonous look-alikes.

WHERE TO FIND: Pacific hemlock-parsley is a coastal species that grows along the upper edge of sandy beaches, on moist coastal bluffs and in tidal marshes and bog woodlands. It is common along the coast of BC, including in Quw'utsun territory.

CULTURAL KNOWLEDGE: This name also pertains to Queen Anne's lace, the introduced wild carrot (*Daucus carota*). Luschiim knew there was an original "wild carrot" native to Quw'utsun territory that people used to eat (June 24, 2011), and his description fits that of Pacific hemlock-parsley:

> **Shewuq**. Long and white, skinny compared to a garden carrot ... much more tapered to the end; [grows] by the beach. The leaves look like carrot, finely divided, and the taste and smell of that is very similar to a modern carrot. But having said that, I was shown another one that had a similar taproot but different leaves, and I was told there's lots of it out on the islands, Orcas Island, around there. (June 24, 2011)

He also confirmed that the original carrot was similar in appearance to a photo of Pacific hemlock-parsley. He said the original wild carrot grew only at the west end of the Malahat reserve. "And that's where I got it, and I brought it to an Elder, and he said, 'Oh, **shuw'qeen**!'" (June 24, 2011).

There is another plant called **shuw'qeen**, which has very similar leaves but "very little root to it. It's like a carrot but only leaves ..." (Luschiim, June 24, 2011).

FIREWEED (*Epilobium angustifolium,* syn. *Chamerion angustifolium*)

ONAGRACEAE (EVENING PRIMROSE FAMILY)

HUL'Q'UMI'NUM' NAME: Not recalled by Luschiim

DESCRIPTION: A tall, often patch-forming perennial herb growing from rhizome-like roots, with erect stems, usually unbranched, often two metres or taller. The stems are smooth and greenish or reddish. The leaves are lance-shaped, smooth-edged or finely toothed, with very short stalks, and arranged alternately along the stalks. The flowers are reddish purple and four-petalled, in dense elongated racemes. The fruits are long, slender capsules splitting longitudinally to release the seeds, each with soft, white parachutes.

WHERE TO FIND: Open woods, thickets, meadows, roadsides, burns and clearings from sea level to upper elevations; abundant throughout BC, including Quw'utsun territory.

CULTURAL KNOWLEDGE: People used to eat the insides of the young shoots of fireweed in the spring. In the old days, too, when they killed a deer, they would clean it right away and some hunters would fill the cavity with fireweed plants and flowers, which helped to soak up the blood and give the flesh a sweet taste. This started the curing/ seasoning process (Luschiim, September 24, 2010, June 24, 2011, and May 25, 2017). Luschiim noted that his family uses maple leaves for this purpose, but fireweed and bracken fern fronds were also used to give the meat a good flavour. The fluff from the fireweed fruits was sometimes woven in with cedar bark, mountain goat wool and/or dog wool to add softness to blankets and clothing; cattail fluff was also used for this purpose (Luschiim, June 24, 2011).

YELLOW AVALANCHE LILY, or YELLOW GLACIER LILY

(*Erythronium grandiflorum*)

LILIACEAE (LILY FAMILY)

HUL'Q'UMI'NUM' NAME: Not recalled by Luschiim, but he noted that there is a Hul'q'umi'num' name for this plant, which is possibly similar to the Mainland Halkomelem name for this lily, ***sk'ámuth***

DESCRIPTION: A perennial herb growing from an elongated, slender bulb, with flowering stems usually 15 to 20 centimetres high. There are two basal leaves, bright green and oblong, up to 20 centimetres long and smooth. There is usually a single flower borne on a smooth, upright stem. The flowers are golden yellow and nodding, each with six recurving tepals, six stamens and a single three-chambered pistil. The fruiting capsules are club-shaped and usually three to four centimetres long, encasing many papery, light brown seeds.

WHERE TO FIND: Moist meadows mostly in montane areas, common throughout the mountains of southern BC, but rare in Quw'utsun territory;

Yellow avalanche lily

small populations grow on both peaks of Mount Prevost, and the species also occurs in the Mount Arrowsmith area.

CULTURAL KNOWLEDGE: Although Luschiim could not recall the name for yellow avalanche lily, he was told by Al Scott Johnny of Nooksack (his cousin) that the bulbs were eaten by the Hul'q'umi'num', as they were by the First Peoples of the southern Interior. Al Scott Johnny had been told this information by Rose August of Westholme, who said there was a name for these bulbs. She is from Cowichan but was married to Johnny August from Westholme (Luschiim, September 24, 2010). Luschiim agreed that the Quw'utsun name would be very similar to that of the people from around Chilliwack: *sk'ámuth*. As well as on Mount Prevost (Swuq'us), the plant grows on Xwaaqw'um (a place on the north side of Cowichan Lake, east of Youbou; the name means "female common merganser," *xwaaqw'*; June 24, 2011). This name is also applied to Burgoyne Bay on Salt Spring Island. The mountain where this creek starts is called Xwaaqw'um Smeent, "merganser mountain" (Luschiim, June 24, 2011, and November 2017).

Luschiim was told that yellow avalanche lily bulbs are similar to *speenhw* (camas) and, like these bulbs, they were pit-cooked before being eaten (June 24, 2011).

SEASIDE STRAWBERRY (*Fragaria chiloensis*); WOODLAND STRAWBERRY (*F. vesca*); and BLUELEAF STRAWBERRY (*F. virginiana*)

ROSACEAE (ROSE FAMILY)

HUL'Q'UMI'NUM' NAME: Woodland strawberry: *st'i'luqw*; blueleaf strawberry and seaside strawberry: *stsi'yu* (also used for garden strawberries)

DESCRIPTION: Wild strawberries are herbaceous perennials that grow from rhizomes and spread by runners, often forming patches. The leaves are long-stalked and three-parted, the leaflets with serrated edges, varying from leathery and shiny (seaside strawberry), to bright green and soft (woodland strawberry), to bluish green (blueleaf strawberry). The flowers are white and five-petalled with yellow centres and many stamens, borne singly or in small clusters. The berries, red when ripe, are soft, sweet and fragrant. The flowers and fruits of seaside and blueleaf strawberries are generally shorter-stalked and closer to the ground; those of woodland strawberry are on longer, more upright stems. The "berries" of the first two are more spherical; those of woodland strawberry are often more elongated. Achenes (one-seeded fruits) are embedded on the fleshy receptacle of the "berry."

WHERE TO FIND: Woodland strawberry and blueleaf strawberry grow in open forests, meadows and clearings from sea level to subalpine areas, whereas seaside strawberry grows in dunes and bluffs along the coastline. All three species occur in Quw'utsun territory.

Blueleaf strawberry fruit

CULTURAL KNOWLEDGE: Luschiim recognizes all three species of wild strawberries, noting that the name *st'i'luqw* refers to the one with "long, thin, dangling berries" (June 24, 2011). However, the names are somewhat interchangeable, as Luschiim described:

> The other one that's up here, the reason why I say "up here" is that there is a similar one with waxy leaves by the sea ... *stsi'yu* is the other one. Now I think both berries that are not dangling are now being called *stsi'yu*. I don't know the other name ... So some of us recognize the difference, not everyone. As far as I know, both types have the same uses, in the way of their leaves and their runners, for making tea to counteract diarrhoea. (June 24, 2011)

Blueleaf strawberry leaves and flower

Luschiim talked further about wild strawberries and how valuable they are:

> The berries are very good for people and for deer and other creatures. For me, when I was hunting all the time in my teens and 20s, I knew where all the berries were. So to me, especially the *st'i'luqw* was an indicator plant for me, 'cause when they're ripe you don't have to see them. You go by your nose; you can smell the berries ... Many times, you start hunting as soon as you can smell the berries. Or your deer or moose or whatever it may be ... whatever you're hunting ... deer, moose or even people. (June 24, 2011)

CHOCOLATE LILY, or RICEROOT (*Fritillaria affinis*) **and** NORTHERN RICEROOT (*F. camschatcensis*)

LILIACEAE (LILY FAMILY)

HUL'Q'UMI'NUM' NAME: Not remembered by Luschiim, possibly similar to the Squamish name, **lhasem**, which Luschiim said means "something slides off," and might refer to the flower hanging down (one of the names listed for tigerlily, **stl'ults'uluqw'us**, page 231, might possibly pertain to northern riceroot; Luschiim was uncertain about this)

DESCRIPTION: Both of these lilies are perennial herbs growing from fleshy bulbs, which are surrounded by numerous small rice-like bulblets. The stalks can grow as tall as 60 to 80 centimetres. The leaves are bright green and grow in one to two whorls of three to five leaves each in chocolate lily, and two to three or more whorls

Chocolate lily, or riceroot (*Fritillaria affinis*)

of fi␣ to ten leaves each in northern riceroot. Chocolate lily flowers are usually borne singly or in a small terminal cluster of two to five. The flowers are brownish purple, usually mottled with greenish yellow, and broadly bell-shaped, each with six oblong to oval tepals. Northern riceroot flowers are usually clustered several together, relatively short-stalked and spreading or somewhat nodding. The flowers are bronze to deep purplish brown, rarely yellowish or mottled, and narrowly bell-shaped with six elliptical, somewhat pointed tepals. The fruits of both species are six-angled capsules, each two to three centimetres long, those of chocolate lily prominently winged at the angles. The capsules split open in three segments when ripe to release numerous papery seeds.

WHERE TO FIND: Chocolate lily grows on dry to medium-dry grassy bluffs, prairies and open forests from the coast to upper elevations. Northern riceroot tends to grow in coastal marshes and estuaries and along ocean shorelines. Both occur in Quw'utsun territory.

CULTURAL KNOWLEDGE: Luschiim recognizes the two different species, one (northern riceroot) growing in salt marshes, and one with flowers with yellow specks in them (chocolate lily), which grows up higher (June 24, 2011). The bulbs of both of these species would have been cooked and eaten by the Quw'utsun, but people haven't used them for some time.

Luschiim noted that northern riceroot grows at Ivy Green Park (a Stz'uminus-run campground called Kwi'kwunluhw, "little root") in Ladysmith Harbour, as well as at Chemainus (September 24, 2010). Chocolate lily, also often called riceroot, is more widespread in Quw'utsun territory, such as on Mount Tzouhalem. Luschiim commented, "They're around there. Not really lots, but they're there—the ones that grow up on the hill, not the ones that grow on the beach: that's a different one [northern riceroot]. They would have dug those [chocolate lily bulbs] as well" (January 23, 2011).

The flowers of both of these are pollinated by carrion flies and smell bad, like rotten meat. Luschiim recalled, "Yes, 'cause as a maybe even pre-teen, I was hunting with my cousin, Waldy Joe, and there's lots up on Mount Tzouhalem [at the] west end. So I picked a handful and I found out they don't smell very good!" (June 24, 2011).

SWEET-SCENTED BEDSTRAW (*Galium triflorum*)

RUBIACEAE (MADDER FAMILY)

HUL'Q'UMI'NUM' NAME: *Hwum'kw'iyathun* or *hwum'kwiiyathun* (also used for large-leaved avens, *Geum macrophyllum*, which has similar burr-like fruits)

DESCRIPTION: A creeping or somewhat climbing perennial herb with leaves in whorls of six (or sometimes fewer), elliptical or lance-shaped and wider toward the tip. The leaves and stems are covered with hooked hairs or bristles. The flowers are usually in sets of three, very small and whitish, with a four-lobed, saucer-shaped corolla. The fruits are small bristly burrs, green at first, turning brown later.

WHERE TO FIND: Moist to medium open woods, streambanks and clearings throughout BC; common in Quw'utsun territory. A larger, more robust species called cleavers, or common bedstraw (*Galium aparine*), is also common but tends to grow along the shoreline or in disturbed areas.

CULTURAL KNOWLEDGE: The name for this plant is also used for large-leaved avens. Although the plants and flowers are very different, both have "stickies," and they grow in similar places—moist woods and streambanks. Luschiim explained the meaning of the name: "Those little burrs attach on to you. So **mukwit** is to see something and pick it up, so when you're walking by [it attaches itself to you]. And you run your hand along there, it's [rough] yeah. So I know this one as **hwum'kwiiyathun'**" (June 24, 2011). Sweet-scented bedstraw also had other, spiritual uses.

LARGE-LEAVED AVENS *(Geum macrophyllum)*
ROSACEAE (ROSE FAMILY)

HUL'Q'UMI'NUM' NAME: *Stsi'yeen'* (also used for sweet-scented bedstraw, *Galium triflorum*, which has similar burr-like fruits)

DESCRIPTION: A perennial herb growing from a short rhizome, with erect stems up to one metre tall, sometimes branching. There are several basal leaves, which are long-stalked and pinnately divided into five to nine segments with a larger heart-shaped, rounded terminal segment. The stem leaves are smaller and alternate, deeply three-lobed. The flowers are quite small, yellow and buttercup-like, each with five petals and borne in a terminal cluster. The fruits are achenes, borne in globe-shaped clusters, each achene with a hooked beak. When ripe, the fruit clusters break apart and the achenes easily hook onto clothing or animal fur.

WHERE TO FIND: Moist open woods and meadows, roadsides, clearings and streambanks throughout BC, and common in Quw'utsun territory.

CULTURAL KNOWLEDGE: Dorothy Charles, originally from Penelekut, then at Nooksack (the sister of Norman Johnny), said that a person could chew the

leaves of this plant and apply them as a poultice for sores, such as on an injury on the shin (Luschiim, January 23, 2011; see also broad-leaved plantain, page 243). Luschiim thought that this was a plant he was told to be "cautious" about using, probably because it looks like buttercups, and these are known to be poisonous and cause blistering and skin sores. He commented, "I'm going back 63, 64 years ago that was being shared with me" (June 24, 2011).

RATTLESNAKE PLANTAIN

(*Goodyera oblongifolia*)
ORCHIDACEAE (ORCHID FAMILY)

HUL'Q'UMI'NUM' NAME: *Shqul'iiqep'nuts* ("attached at the bottom to the ground"; Luschiim, June 24, 2011). There are other names for this orchid, but they are considered sacred and used only at ceremonial times by those with the rights to use them, so are not included here.

DESCRIPTION: A low-growing herbaceous perennial growing from thick rhizomes, each separate plant having spreading, oval basal leaves that are pointed, smooth and distinctively marked with white spots. Flowering occurs in the summer, and usually only a few plants within a patch produce flowers. The flowers are borne on a single upright stalk and are small and greenish white, arranged in a terminal spike. The fruits are brown capsules containing many tiny seeds.

WHERE TO FIND: Open woods and forest edges, often in mossy places; found throughout British Columbia and sporadically in Quw'utsun territory.

CULTURAL KNOWLEDGE:
Luschiim explained that the name, *shqul'iiqep'nuts*, refers to *sqep'* ("attached to something") and *nuts* ("bottom end"). This is because the plants are lightly attached to the ground, "So when you go to get them, the roots are barely under the moss" (Luschiim, June 24, 2011). This plant has important spiritual applications and must be harvested carefully so as not to deplete the populations.

OREGON GUMWEED (*Grindelia stricta*)
ASTERACEAE (ASTER FAMILY)

HUL'Q'UMI'NUM' NAME: Not recalled by Luschiim

DESCRIPTION: A perennial herb growing from a taproot with erect stems, usually branching, smooth or with long, soft hairs. The basal leaves are alternate, elongated, toothed or smooth-edged, and the lower ones are wider toward the top. The flower heads are solitary or in a flat-topped cluster, daisy-like with 10 to 35 bright yellow ray flowers per head and sticky whitish bracts around the base of the head, the outer bracts long, slender and spreading.

WHERE TO FIND: This plant grows mainly along the coast at the upper ends of beaches, in tidal marshes, on rocky bluffs and along roadsides; common in coastal BC and coastal areas of Quw'utsun territory.

CULTURAL KNOWLEDGE: Luschiim knew this plant well and commented,

> It's always sticky, all the time ... It's a hand cleaner. You just go like that with the leaves [rub them] and the gummy part [is a] de-scentizer. Clams, when you're clam digging, you don't notice the smell, but other people do, so they use that to de-scent. It's out almost all year around. Even when they're gone, there's new shoots coming out. You can get it all year round. So [it's known] for the scentizer, the scent, the smell, and I'm told you can get it any time of the year.
> (June 24, 2011)

GRASSES and
GRASS-LIKE PLANTS (many species)
POACEAE (GRASS FAMILY)

HUL'Q'UMI'NUM' NAME: General name for grasses, hay and grass-like plants: **saaxwul**; foxtail barley, or speargrass (*Hordeum jubatum*): **teeqe'lh**; American dunegrass, or beach wildrye (*Leymus mollis*): **slhukw'aa'iy** [uncertain]; common rush (*Juncus effusus*): **p-shey'** ("sharp grass"); grassy field or meadow (general term): **spulhxun**

DESCRIPTION: There are many types of grasses in Quw'utsun territory, both native and introduced. They are similar in having long, narrow, pointed leaves; small, wind-pollinated flowers that are borne in clusters; and fruits that are grains, small and rounded or elongated. There are other grass-like plants, including sedges (*Carex* spp. and related species) and rushes (*Juncus* spp. and related species), which might also be called by the same general Hul'q'umi'num' name for grasses.

American dunegrass, or beach wildrye (*Leymus mollis*)

WHERE TO FIND: Grasses are common in a whole range of habitats throughout Quw'utsun territory, from coastal areas to mountaintops and from very wet to very dry sites.

CULTURAL KNOWLEDGE: Luschiim commented that there are several types of native grasses up at Eagle Heights. Grasses are known and used generally for laying food on or for other domestic purposes. Basket-makers harvest the young stems of some grasses, such as reed canary grass (*Phalaris arundinacea*), cure them, then split them and use them, along with bitter cherry bark (***t'ulum'***), to decorate split coiled cedar-root baskets. Luschiim was told that this grass was "brought in," but is not sure whether or not it is introduced (April 16, 2016).

Foxtail barley (speargrass) and other grasses with sharp, pointed awns have particular importance because of their potential harm. Luschiim recalled, "So right from early childhood, 1946, '45, '47, we were really cautioned that we don't get any of that [those sharp grass fruits] in with our berries. Because that was part of our everyday; one of the children's responsibilities was to pick berries ..." (June 24, 2011). These fruits and the sharp, black seeds of sweet cicely (*Osmorhiza chilensis, O. purpurea*) can cause choking and can become embedded in the throat. Luschiim said, "That's what we were told: 'It can kill you, kill someone!'" (June 24, 2011).

Luschiim said that the native American dunegrass, which grows on dunes and the upper edge of beaches all along the coast, is being replaced by introduced beach grass, and it is hard to find the native grass anymore. He said it was formerly used for weaving "by those who were really good weavers" (June 24, 2011).

When asked about the plant called ***p-shey'***, Luschiim said it was definitely not the one often referred to as "cut-grass"—small-flowered bulrush (*Scirpus microcarpus*)—or slough sedge (*Carex obnupta*). Instead, he described ***p-shey'*** as having round, dark green stems, growing about a metre high, and fitting the description of common rush (*Juncus effusus*). He noted that there was a lot growing all around the area at the corner of Allenby and Koksilah Roads, across the bridge and past the band office, where we found it growing. He said that it has spiritual uses (April 18, 2016, and November 2017).

COW-PARSNIP (*Heracleum maximum,* syn. *Heracleum lanatum*)

APIACEAE (CELERY FAMILY)

HUL'Q'UMI'NUM' NAME: Non-flowering leafy plants or leaf stalks: ***yaala'*** (Luschiim, June 24, 2011); flowering plants (or plants with one stalk and one leaf): ***saaqw'*** (Luschiim, June 24, 2011)

DESCRIPTION: A robust herbaceous perennial growing up to three metres tall from a dense, branching taproot, with large, three-parted basal leaves with long, hollow stalks and hollow, upright flowering stems. The stem and leaves are covered with dense hairs. The flowers are small and white, borne in large, flat-topped, umbrella-like clusters, up to 20 centimetres across. The fruits are heart-shaped, flattened and winged.

WHERE TO FIND: Moist meadows, streambanks, roadsides and edges of forest, from lowlands to alpine areas; common throughout BC, including in many parts of Quw'utsun territory.

NOTE: This plant contains phototoxins: compounds called furanocoumarins, which cause extensive skin irritation, including blistering and discolouration of the skin when exposed to sunlight (ultraviolet light). The leaf stalks and young bud-stalks are edible, but only when peeled, and the mature stems should not be eaten.

Cow-parsnip flowers

Cow-parsnip flower buds

CULTURAL KNOWLEDGE: (See note above.) Holland Creek and its estuary—located near Ladysmith—are called Hwsaaqw'um after the edible bud-stalks of the cow-parsnip. Both the leaf-stalks and the bud-stalks of the young plants are edible, but only after being peeled, and only in the spring when the plants are still young and before they flower. Once peeled, the stalks are sweet and celery-like; usually they are just eaten raw. Luschiim noted on April 18, 2016, that the ***yaala'*** was just ready to eat at that time in the Sooke area, whereas on Mount Arrowsmith, he found some that was still ready to eat at the end of the first week of October.

Luschiim considers this plant to be "a cousin to the coltsfoot [*Petasites frigidus*]." He said that people were cautioned to tell the difference between these plants; coltsfoot is a look-alike, but not considered edible (September 24, 2010, and June 24, 2011). He had never heard of the flowering and non-flowering cow-parsnip plants being called "male" and "female," respectively.

BEACH PEA *(Lathyrus japonicus)*
and PURPLE PEA *(L. nevadensis)*

FABACEAE (PEA FAMILY)

HUL'Q'UMI'NUM' NAME: *Tl'ikw'un'* or *tl'ikw'um* (also used for giant vetch, *Vicia nigricans*, and recently for garden peas, *Pisum sativum*; beach pea: *tl'ikw'um ni' 'utu tsetsuw*, "peas at the beach"; purple pea: *tl'ikw'um ni' 'utu tsa'luqw*, "peas up the hill")

DESCRIPTION: These are both perennial herbs growing from creeping rhizomes, forming patches or climbing on other vegetation. The leaves are pinnately compound with tendrils at the ends. Those of beach pea are bluish green and somewhat fleshy, whereas purple pea leaves are thinner. The flowers are reddish purple and similar to pea flowers, growing in fairly dense

Beach pea *(Lathyrus japonicus)*

clusters of usually six to eight. The fruits are pea-like pods that split open when ripe to reveal small round "peas."

WHERE TO FIND: Beach pea grows in sand and gravel above the upper tide line along the coast; purple pea grows in open woods, forest edges and clearings; both occur along the coast of BC and in Quw'utsun territory. Purple pea extends inland across the southern part of the province.

CULTURAL KNOWLEDGE: The "peas" (and pods) of these two species were sometimes eaten, along with those of giant vetch. Luschiim commented,

> Edible pods? Yeah, there's several different kinds [of wild peas and vetches] and some there's not worth bothering with. But the ones I know, they're the biggest pods down there in the wild, and that's called *tl'ikw'un'*, and that name was transferred to the modern-day peas ... I just know they were edible. They're three kinds ... And they're sweet and edible. So there's a bunch of them growing at Hwkw'a'luhwum [Qualicum River], Shts'um'inus [Chemainus] and other places. (June 24, 2011, and November 2017)

Luschiim wouldn't bother to eat these most of the time, unless they were really sweet and tender (April 16, 2015).

TIGERLILY *(Lilium columbianum)*

LILIACEAE (LILY FAMILY)

HUL'Q'UMI'NUM' NAME: *Stl'ults'uluqw'us* or possibly *slhaqwelukw*

DESCRIPTION: An herbaceous perennial growing from a large spherical scaly bulb. The flowering stems are upright with whorls of lance-shaped, pointed leaves spaced along the stalk and large showy orange flowers borne singly or in few-flowered clusters at the top. The tepals (petals and sepals together) are bright orange, speckled with black inside and strongly recurved at maturity. The fruits are club-like, three-parted capsules that split open when ripe to reveal light brown, papery seeds.

WHERE TO FIND: Open woods, prairies and forest edges from sea level to subalpine areas; common across southern BC and occurring sporadically in Quw'utsun territory. Luschiim noted that they are very common in places that recently have been logged.

CULTURAL KNOWLEDGE: The bulbs of this lily are edible but somewhat bitter. Luschiim has eaten them, but said they are an acquired taste (September 24, 2010). Luschiim did not recognize the name **slhaqwelukw**, but said it seems to mean "right by the water." The name may pertain to northern riceroot (see page 219).

BARESTEM DESERT-PARSLEY, or INDIAN CONSUMPTION PLANT (*Lomatium nudicaule*)

APIACEAE (CELERY FAMILY)

HUL'Q'UMI'NUM' NAME: *Q'uxmin*

DESCRIPTION: Perennial herb growing from a taproot, with single to several stems growing to 80 centimetres or taller. The leaves are mostly basal, with stalks, and compound, one to three times divided, with oval-shaped, smooth, bluish-green leaflets. The small, light yellow flowers are arranged in uneven umbrella-like clusters. The fruits are elliptical and ribbed or striped, with distinctive wings running along the length.

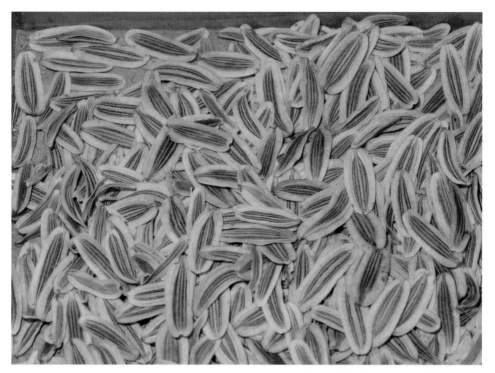

Seeds from barestem desert-parsley, or Indian consumption plant (*Lomatium nudicaule*)

WHERE TO FIND: Dry grassy slopes and rocky bluffs, often near the ocean; open prairies and oak savannahs. Locally common on southeastern Vancouver Island, extending into the southern Interior of BC; growing in several locations in Quw'utsun territory.

CULTURAL KNOWLEDGE: There is a place on the Cowichan River called Hwq'uxminum, "place of *q'uxmin*." The aromatic seeds of this plant are used as a ceremonial incense in the First Salmon Ceremony and for smudging and purification. They are also known as a good medicine for coughs and colds. Luschiim had heard of other people eating the young leaves as greens but said that the Quw'utsun do not eat them (April 16, 2015).

SKUNK-CABBAGE (*Lysichiton americanus*)
ARACEAE (ARUM FAMILY)

Skunk-cabbage flowers

HUL'Q'UMI'NUM' NAME: *Ts'a'kw'a'*

DESCRIPTION: An herbaceous perennial with large, distinctive, bright green oval leaves, one metre or longer, growing in clumps from thick whitish rhizomes. The flowers, which emerge in the spring before the leaves have expanded, consist of a showy, bright yellow sheath surrounding a stalked, club-like spike of densely packed, greenish flowers. The fruits, ripening on the spike, are egg-shaped, each containing one or two seeds.

WHERE TO FIND: Swamps, stream edges and ponds from lowland to montane zones, across British Columbia and frequent in Quw'utsun territory.

CULTURAL KNOWLEDGE: The name *ts'a'kw'a'* is similar to the name for spiny wood fern (*Dryopteris expansa*) in some other Salishan languages, such as Mainland Halkomelem, Squamish and Klallam. The broad, waxy leaves of skunk-cabbage were formerly used as a surface for drying berries and laying food on.

Luschiim told a story about the Beaver People and skunk-cabbage leaves:

A long time ago there was a man poling up the river, the Cowichan River. He was getting close to Cowichan Lake. He was poling, poling his canoe ... then he hears in the distance, at the left of the river, facing upstream, somebody *qwi'qwal'*, so he starts to sneak with this pole.

Quietly poling up the river, he beaches his canoe very quietly; he's listening … It sounds … like people gathered together … He listens, yeah: "'*A 'a siiem' nu siiye'yu, siiem' nu s-ts'lhhwulmuhw sne, sne kwthu hun'tsew*": "My dear relatives, the name you are going to hear, the name that's coming down, has not been used for a long time." And a guy would say the name … Kept going on: "'*A 'a siiem' nu siiye', siiem' nu s-ts'lhhwulmuhw sne, sne kwu ts'elhum 'ut 'uhw. Sne, sne, mich hwiwul siem.*" It would be different each time … The name, you will hear somebody's using it, call a person up, that's using that name … He's sneaking over, peeking through the trees. He is trying to recognize who they are. Their names are going on, lots of people there, by the edge of the lake or pond. And then he wants to get closer because he wants to see who they are. He steps on a branch, [cracking sound]. The branch breaks. The people stop and jump in the water, they come up. And they were beavers! They had lots of goods piled up along the shore, like part of the gifts. All those gifts turn to *ts'a'kw'a'* [skunk-cabbage]. And they were people—people like us. When they dove into the water, they were people. When they come up, they were beavers! So there was beavers, doing their naming with their families … And that was their *ts'a'kw'a'*. That was their gifts … *ts'a'kw'a'* leaves piled up along the shore … That's the short version of that story. (April 16, 2015)

Luschiim mentioned to Liz Hammond-Kaarremaa that his great-grandfather Luschiim told him, when he was three years old, in 1945, that when the black bears and grizzlies wake up in spring after their winter sleep, they gorge on *ts'a'kw'a'* (skunk-cabbage leaves)—"all they can eat"—and run through bushes. This flushes out all the worms that they might have. The bears also eat a certain type of clay, *suy'q'*, before they go into their winter sleep, to kill any worms.

FIELD MINT *(Mentha arvensis)*

LAMIACEAE (MINT FAMILY)

HUL'Q'UMI'NUM' NAME: *Stul'a'tuxw* (see ***sti'a'lus***, "leaning on something"; this mint grows lying on the ground or leaning on other plants.)

DESCRIPTION: A peppermint-scented herbaceous perennial, growing from creeping rhizomes and forming patches. The stems are upright to creeping and square in cross-section with opposite, bright green leaves that are oval and pointed with serrated edges. The flowers are small and mauve, clustered at the leaf axils, and the fruits are small brownish capsules.

WHERE TO FIND: Wet meadows, marshes, ditches and estuaries, growing with sedges and other wetland plants.

CULTURAL KNOWLEDGE: Luschiim talked about a kind of mint growing at Beacon Hill Park, apparently this species. It is a tall mint, growing in a wet area just at the toe of the hill before it drops onto the beach, west of where the little shack used to be, just south of Beacon Hill. Luschiim said there were also some of these mint plants at the creek at Isabel Point, the second creek at a park in Cowichan Bay. He said it grows quite thick there: "lots standing together." There is also mint with purplish flowers growing all over the beach at Fort Rupert in damp, shady areas. It doesn't have an offensive smell; it smells more like mint gum or an air freshener. He said people used to keep small clumps of this mint in a house or room, for spiritual purposes (September 24, 2010).

Felix Jack from Mayne Island told Luschiim about a sweet-smelling mint, probably this species, growing at the second creek from Isabella Point, going home. He said there is a medicine growing there: "You go there, you'll know; nothing else." It was 30 years later before Luschiim went and stopped there and found it. Luschiim also found this mint growing at Wildwood, Yellow Point, beside Merv Wilkinson's barn. He noted that **stul'a'tuxw** is known to be a favoured elk's food.

INDIAN-PIPE (*Monotropa uniflora*)
and CORALROOT (*Corallorhiza maculata*)

ERICACEAE (HEATHER FAMILY): INDIAN-PIPE
ORCHIDACEAE (ORCHID FAMILY): CORALROOT

HUL'Q'UMI'NUM' NAME: The W̱SÁNEĆ people call Indian-pipe "wolf's urine,"
but the plant Luschiim knows by the name ***suxwa' 'utl' stqeeye'***, "wolf's urine,"
is different (it may be coralroot or a similar species).

DESCRIPTION: Both of these flowering plants lack green leaves. Indian-pipe
is parasitic on mycorrhizal fungi associated with trees, and coralroot is a
saprophytic herb growing from coral-like rhizomes. Both grow in clumps
and have upright stems with small, scaly leaves. Indian-pipe, which grows
up to 30 centimetres tall but is usually much shorter, is white when at its
young and flowering stage, blackening with age. Coralroot is deep maroon,
its shoots similar in shape to a drawing Luschiim made of the plant he calls
suxwa' 'utl' stqeeye'. At maturity the stems can grow to 50 centimetres or
higher. Indian-pipe stems are white and waxy looking, terminating in a single,

Indian-pipe (*Monotropa uniflora*)

nodding flower, and overall resembling a pipe. Coralroot has multiple (10 to 40) miniature "orchid" flowers clustered at the top of the stem, each with five spreading petals and a white, purple-spotted (plain or striped in other species) lip at the base. Indian-pipe stems turn upright at fruiting, producing globe-shaped capsules. Coralroot produces egg-shaped, drooping capsules.

WHERE TO FIND: Indian-pipe grows in open, somewhat shaded woods, mainly under coniferous trees, and occurs across British Columbia and in Quw'utsun territory. Coralroot species also occur in moist forested areas, including in subalpine forests.

CULTURAL KNOWLEDGE: Luschiim drew a sketch of the plant he knows as *suxwa''utl' stqeeye'*. He said it is maroon in colour (which fits the description of coralroot) and unbranched. He said it smells really strong—"like a diaper has been folded up for several days. It's really strong." He was bathed in a solution of this plant when he was born (the end of September) and his grandson, born in early April, was also bathed in this solution. He said it grows in higher mountains, including local mountains, but also grows down in the valley, including down by the Bighouse.

Coralroot (*Corallorhiza maculata*)

YELLOW PONDLILY (*Nuphar polysepala*)

NYMPHAEACEAE (WATERLILY FAMILY)

HUL'Q'UMI'NUM' NAME: Possibly **qw'emétxw** (the Upriver Halkomelem name is **qw'emét**, "pull it up by the roots")

DESCRIPTION: Perennial aquatic plant growing from a large, thick rhizome. The fleshy stems, up to two metres long, are submerged in water or partially immersed, depending on the depth of the water. The leaves are large, with long stalks and heart-shaped, bright green, leathery blades. The flowers are large, waxy and cup-shaped, borne on long stalks growing from the rhizome. The stamens are large, yellow and petal-like, and the flower centre consists of a large, bright yellow flattened stigma surrounded by yellow or reddish stamens. The fruits are large oval, leathery green floating capsules with numerous seeds in a jelly-like mass.

WHERE TO FIND: Shallow ponds, lake edges and slow streams and bogs, from sea level to montane ponds and lakes.

CULTURAL KNOWLEDGE: Luschiim did not recall a name for this aquatic plant but had heard that one can use the rhizomes ("roots") as an emergency food, although they must be boiled in at least four to six changes of water, and in "the biggest pot you can get," or they are too bitter to eat (September 24, 2010, and May 25, 2017). He said this

plant grows across the Fraser River from Tl'uqtinus, at the south end of the bridge, along with wapato (**sqewth**). He said there is a bit of a walkway there; you go through farm fields to get to the south side of the bridge, and just before that is a little walking trail and a little road that goes into the water. You could look across and see yellow pondlilies (April 18, 2016).

BRITTLE PRICKLYPEAR CACTUS (*Opuntia fragilis*)

CACTACEAE (CACTUS FAMILY)

HUL'Q'UMI'NUM' NAME: *Thuthuhw* (the last part of this word means "it disappears" or "hidden")

DESCRIPTION: A perennial herb with succulent, prostrate stems forming dense mats, light green in colour. The stem segments are rounded, covered with spiny bristles and straight light brown spines that are up to three centimetres long. The flowers are large (three to five centimetres across) and single, with thin yellow petals and numerous stamens. The fruits, up to two centimetres long, are fleshy, spiny and somewhat pear-shaped.

WHERE TO FIND: Dry gravelly or rocky slopes across southern British Columbia, but rare in Quw'utsun territory, on southeastern Vancouver Island and the Gulf Islands.

CULTURAL KNOWLEDGE: Luschiim said that this plant was used for food and spiritual purposes. As food, you can roast it, knock the prickles off and eat it, although Luschiim himself hasn't tried it (September 24, 2010).

COLTSFOOT *(Petasites frigidus* var. *palmatus)*

ASTERACEAE (ASTER FAMILY)

HUL'Q'UMI'NUM' NAME: *Sq'une'yux* (Luschiim, June 24, 2011)

DESCRIPTION: A herbaceous perennial that grows from creeping roots and forms patches. The stems are erect, covered in white woolly hairs, up to 50 centimetres tall but usually shorter; the flowering stems appear before the leaves. The basal leaves are large and long-stalked, palmately lobed and often hairy or woolly; the stem leaves are much reduced. The flower heads are whitish or pinkish, each with numerous disk flowers, and form rounded clusters. The achenes are small, with a pappus of white hair–like bristles.

WHERE TO FIND: Wet ground, in ditches, along streambanks and seepage areas, from lowlands to alpine ones; common through BC, including in Quw'utsun territory.

CULTURAL KNOWLEDGE: Luschiim said this plant is used for spiritual and possibly medicinal purposes, and that it is not good for people. Its name pertains to tasting bad. He noted that it is similar to *yaala'* (cow-parsnip) but is not edible (September 24, 2010, and April 18, 2016).

BROAD-LEAVED PLANTAIN (*Plantago major*)

PLANTAGINACEAE (PLANTAIN FAMILY)

HUL'Q'UMI'NUM' NAME: *Sxu'enhween*

DESCRIPTION: A perennial herb growing from a fibrous root. The basal leaves are broadly elliptical, tapering into a long stalk, smooth-edged or slightly toothed and strongly and distinctively parallel-veined. The flowers are small and greenish, borne on a long, thin spike up to 30 centimetres long, and the fruits are brown, egg-shaped capsules containing few to many seeds.

WHERE TO FIND: Medium to dry areas in disturbed soil, roadsides, fields and waste places from sea level to upper areas; common in southwestern BC, including Quw'utsun territory.

CULTURAL KNOWLEDGE: Although this plant is said to have been introduced from Europe, Luschiim had been told by Quw'utsun Elders that it was always here (April 18, 2016). (Note that there may well be indigenous and introduced strains of this species.) It is well known as a medicinal plant with multiple applications, including for treating broken bones. A person would chew the leaves and apply them as a poultice for sores, such as on the bony shin part of the leg when an injury would not heal (Luschiim, January 23, 2011).

DOCK, or "COFFEE PLANT" species, including BITTER DOCK (*Rumex obtusifolius*); YELLOW DOCK (*R. crispus*); WESTERN DOCK (*R. aquaticus*); and other *Rumex* spp.

POLYGONACEAE (KNOTWEED FAMILY)

HUL'Q'UMI'NUM' NAME: *T'umasu* ("birthing medicine")

DESCRIPTION: Perennial herbs growing from stout taproots, with erect stems, up to one metre tall. The leaves vary with the species. The leaf blades of

yellow dock are wedge-shaped and pointed, those of bitter dock are oval and heart-shaped at the base, and those of western dock are elongated and pointed. The flowers of all these species are numerous, small and greenish, in elongated, often branching clusters, mostly at the ends of the stalks. They ripen to brown fruits that collectively resemble coffee grounds on a stem.

WHERE TO FIND: Various dock species are common in open areas, disturbed ground, old pastures and road edges. Western dock, which is native to British Columbia, grows in tidal flats and river estuaries.

Western dock leaf (*Rumex aquaticus*)

Western dock (*Rumex aquaticus*)

CULTURAL KNOWLEDGE: Luschiim noted that there were various types of dock, some with curly leaves, some with flat leaves. He was told that some had tubers but hasn't seen these. He said they were used medicinally but did not know the details (April 18, 2016). The Hul'q'umi'num' name for dock species, *t'umasu*, may also be applied to cultivated rhubarb, but although young dock leaves can be cooked and eaten, one should never eat rhubarb leaves. Luschiim said that children also used the brown grainy fruits for play tea or coffee (November 2017).

WAPATO (*Sagittaria latifolia*)

ALISMATACEAE (WATER-PLANTAIN FAMILY)

HUL'Q'UMI'NUM' NAME: *Sqewth* (also used for garden potatoes)

DESCRIPTION: An aquatic or semi-aquatic perennial herb growing from tuber-producing rhizomes, with stems growing to 50 centimetres or taller (see photo of camas bulbs and wapato tubers on page 203). The leaves that extend above the surface of the water are long-stalked with arrowhead-shaped blades, whereas submerged leaves are lance-shaped or ribbon-like. The flowers are small and white, three-petalled and arranged in an open cluster. Male and female flowers are borne separately. The fruits are winged achenes in a globular cluster.

WHERE TO FIND: Grows in wet ponds, lakeshores, marshes and ditches from lowlands to montane areas; formerly very common in the Fraser Valley and Fraser River delta; uncommon on Vancouver Island.

CULTURAL KNOWLEDGE: When Luschiim was visiting the site of the Quw'utsun village of Tl'uqtinus on Lulu Island, along the south arm of the Fraser River, he saw some *sqewth* plants across from the old village site. He described how there is a bit of a walkway, and you can walk along through farm fields to get to the south side of the bridge. Just before you get to the bridge, there is a little road that goes down into the water. Looking across from there, he saw both yellow pondlilies and *sqewth* (April 18, 2016). His mother, Violet Charlie, who was ninety-one at the time she told him, said that there had been some wapato growing in a lake on an island near ("this

side of": west of) Sidney. Violet, when she was under 10 years old, used to go with her own mother to sell Cowichan sweaters at Sidney. On their way home, they stopped at this island. They went up a little hill, and there was a swamp or pond where they used to get *sqewth*. This was one of the only places where people could harvest it, aside from the Fraser Valley. Luschiim has eaten these "potatoes" but only a small amount (April 18, 2016).

AMERICAN GLASSWORT, or SEA ASPARAGUS (*Sarcocornia pacifica*, syn. *Salicornia virginica*)

AMARANTHACEAE (AMARANTH FAMILY)

HUL'Q'UMI'NUM' NAME: *Shxwiil'nuts* or *sxwiil'nuts* (versions of this name are applied to arrow-grass, *Triglochin maritima*, in the Shíshálh language and other Salish languages)

DESCRIPTION: American glasswort is a perennial herb, growing from a branching rhizome, with prostrate stems, forming dense mats in the upper intertidal zone and tidal marshes. The leaves are small and scale-like. The bluish-green, waxy-looking stems are succulent, jointed and branching, bearing small flowers in groups of three sunken into the tips. The fruits are single-seeded membranous bladders enclosed by scales.

WHERE TO FIND: Tidal marshes and beaches; common along the BC coast, including along the shoreline in Quw'utsun' territory.

CULTURAL KNOWLEDGE: Luschiim described the plants he calls *sxwiil'nuts* or *shxwiil'nuts*:

> They're just the size of an average pencil, brown, long. They're very salty ... green. They're good in the spring but you can ... [find them up until] October. They're a little bit stringy. But you can eat them ... got a bunch of branches ... there's lots of them at the mud flats at the estuary of the Cowichan River ... (December 7, 2010)

He said that Ronnie Alphonse had eaten these but he himself had not eaten them. (See also under seaside arrow-grass, page 251.)

TULE, or ROUNDSTEM BULRUSH (*Schoenoplectus acutus*)

CYPERACEAE (SEDGE FAMILY)

HUL'Q'UMI'NUM' NAME: *Wool'* (Luschiim, December 2010)

DESCRIPTION: Perennial wetland herb, growing from thick rhizomes and forming large stands. The stems are cylindrical and firm, with a hard pith centre, tapering toward the top. The leaves are sheathed, basal and poorly developed; the flowers are small and petal-less, concealed by overlapping scales forming spikelets clustered at the top. The fruits are small achenes.

WHERE TO FIND: Lakeshores, old lagoons, marshes and stream edges from lowlands to montane areas; common throughout BC, including in Quw'utsun territory (for example, on Lake Cowichan).

CULTURAL KNOWLEDGE: The round, firm but lightweight stems were formerly harvested from Cowichan Lake and other wet areas and sewn into large mats. Luschiim noted that there are special places where people gathered tule and cattail; the stems and leaves vary in strength and texture. Some of them, if they are used to make mats, will fall apart very quickly. This is called *thuphwum*, "falls apart"—a word used for the mats that don't stand very much wear because the stems (tule) or leaves (cattail) are brittle. Long needles of *q'eythulhp* (oceanspray) were used to sew the mats together using twine from stinging nettle or other fibre.

Koksilah Ridge, or Eagle Heights, is called Hwsalu'utsum', "place with bulrush or tule mat shelters" (Rozen, 1985, #252).

BROAD-LEAVED STONECROP (*Sedum spathulifolium*) **and** SPREADING STONECROP (*S. divergens*)

CRASSULACEAE (ORPINE FAMILY)

HUL'Q'UMI'NUM' NAME: Not recalled by Luschiim

DESCRIPTION: These are low-growing succulent plants of rocky bluffs, forming dense patches. The leaves of *S. spathulifolium* are flattened and spoon-shaped, bluish green in colour, turning to red; those of *S. divergens* are more oval and lighter green, tinged with red or entirely reddish. The flowers are bright yellow, growing in flat-topped clusters at the ends of elongated stems. The fruits are clustered, spreading follicles, few in number.

WHERE TO FIND: Rocky bluffs, cliffs and talus slopes from coastline to montane areas. Both species occur in southwest British Columbia, although *S. spathulifolium* is restricted to Vancouver Island and the Gulf Islands; both occur in Quw'utsun territory.

Stonecrop (*Sedum divergens*)

CULTURAL KNOWLEDGE:

Luschiim did not remember the name for stonecrop, but recognized that there are at least two different kinds, and that one (*S. divergens*) was a really colourful red. He said it was a good food. Both kinds were good medicine for burns and other injuries and were also spiritual medicines. He noted that there is lots of the red stonecrop (*S. divergens*) on the mountaintop of Mount Whymper, called Muq'meqe', or "Snowy owl" (April 18, 2016).

SEASIDE ARROW-GRASS

(*Triglochin maritima*)

JUNCAGINACEAE (ARROW-GRASS FAMILY)

HUL'Q'UMI'NUM' NAME: Some people call this plant *sxwillnuts*, but Luschiim applies this name to American glasswort, *Sarcocornia pacifica* (see page 247).

DESCRIPTION: A bright green, grass-like, somewhat succulent perennial herb growing from rhizomes, with stems growing to one metre or taller. The leaves are basal, tufted, narrow and elongated, and the flowers are green and small, borne in a spike at the end of a long stem. The fruits are small and egg-shaped, splitting into six divisions, with one seed each.

WHERE TO FIND: This grass-like plant is common in tidal marshes, shorelines and wet meadows, mostly along the coastline in Quw'utsun territory. It grows throughout southern BC and can occur in montane areas in the Interior.

CULTURAL KNOWLEDGE: Luschiim noted that seaside plantain (*Plantago maritima*) is similar to seaside arrow-grass, although he did not know of a name for that species. He would apply the name *sxwillnuts* to American glasswort, not arrow-grass (September 24, 2010). He had not heard of the name "sea onions," sometimes used by Shíshálh and other Coast Salish people for arrow-grass. One can eat the young plants in the spring, but the mature ones are considered poisonous. (**NOTE:** They are indeed toxic, containing cyanide-producing compounds in mature and flowering plants.)

Luschiim was shown this plant as *shxwiilnuts* by more than two people and said that there is lots growing at Cowichan Bay and lots in Lummi: "any beach with a flat place." The young plants are eaten by lots of people today: "everyone would get it." The older plants are too tough, however. Luschiim noted that when they were travelling by canoe (and later by motorboat) and needed food, they could harvest this vegetable on the shores of Narvaez Bay or on Tumbo and Cabbage Islands. However, he warned that today, many places are too contaminated to harvest food from. You have to watch for oil in the sea (April 18, 2016).

CATTAIL (*Typha latifolia*)
TYPHACEAE (CATTAIL FAMILY)

HUL'Q'UMI'NUM' NAME: Plant: **stth'e'qun** (literally "something with hair on the top"); fibrous edge of the leaf: **slhup-nuts**

DESCRIPTION: A robust perennial herb growing from an extensive network of rhizomes and strongly patch-forming. The stems are up to three metres tall, erect, cylindrical, pithy and hard. The leaves grow in clusters, sheathing around the youngest ones, and growing to two metres or longer, the blades flat and up to two centimetres wide, slightly pointed and spongy or pithy. The flowers are crowded into a dense, terminal spike; the yellow, male (pollen-bearing) flowers are at the top, and the brown, female part of the spike—the "cat's tail"—is below, each portion usually 10 to 20 centimetres long. The fruits are small dry follicles, with seeds attached to a fluffy "parachute," which, when the spikes are fully mature, are released to blow away on the wind.

WHERE TO FIND: Wetlands, ditches, ponds and lake shores from lowland to montane areas; common in southern British Columbia, including in Quw'utsun territory.

CULTURAL KNOWLEDGE:
Luschiim noted that there are two kinds of reeds: one, tule, is called **wool'** and the other one, cattail, **stth'e'qun**:

Your **stth'e'qun**, that's your flat leaves. There's an edge on that flat stem. You grab that edge, and you pull it off. It's like a twine. That's your **slhup-nuts**. That was used for holding things together, whenever you needed strong material. You could even twine it into a heavier twine and keep on adding to it. (December 16, 2010)

Earlier, he had mentioned that "when you're doing reed mats, got your thread right there; cattail mats, sew or bind" (April 18, 2016).

The long, somewhat spongy leaves of cattail were used, like tule stems, to make large mats that were formerly used for mattresses, room dividers, coverings for summer shelters, sails for canoes and as general-purpose mats. Formerly, the leaves or stems were often sewn together, using long wooden needles: "When they're making reed mats today, they are kind of woven; in the past you had a long needle and you poked them; it was called **tth'qw'e'lhtun**—the long needle for sewing reed mats [about a metre long]" (April 18, 2016).

These needles were usually made of oceanspray (**q'eythulhp**). Luschiim said one could also use mock orange, but that wood is not very plentiful, and you could use it for other important tools or implements (April 18, 2016).

Luschiim noted that some cattail leaves and tule stems are stronger and more durable than others, able to withstand rigorous use, whereas others are too brittle, and mats made from them would fall apart easily. He commented that families pass down knowledge about where to harvest good-quality cattails and cedar for weaving (see page 72). He continued:

One of the places we went, my family, on the Luschiim side, is a place called Marietta, by Lummi Island, near Bellingham. That's where we went for reeds that were not **thuphwum**. So I visited the place, just to have a look at those reeds—lots of them there. (December 16, 2010)

Luschiim said that people would make tent-like structures from reed mats, but wasn't sure whether these were of cattail, or tule, or both kinds (June 24, 2011). The seed fluff from the ripened heads of cattails was used for insulation and as a fire-starting material (November 2017).

STINGING NETTLE *(Urtica dioica)*

URTICACEAE (NETTLE FAMILY)

HUL'Q'UMI'NUM' NAME: *Tth'uxtth'ux* (derived from "poison" or "stinging")

DESCRIPTION: A perennial herb growing from strong rhizomes and often forming dense patches. The stems are upright, usually unbranched, and can be one to three metres high. The leaves grow in opposite pairs from nodes along the stem, stalked, with the blades varying from narrowly lance-shaped to oval, sharply pointed with coarse, pointed teeth. The flowers are small and green, borne in small clusters in the upper leaf axils. The fruits are small oval achenes.

WHERE TO FIND: Moist to medium open woods, forest edges, clearings and floodplains from sea level to lower montane areas. Found throughout BC and particularly common in old First Nations village sites. There are both native and European populations of this plant.

CULTURAL KNOWLEDGE: This distinctive plant is a food, medicine, fibre and spiritual plant. Luschiim noted that wherever there are settlements, including midden sites, nettles grow really well. The plants must be cooked before being eaten, and only the young plants, just a few inches high, should be eaten.

The fibres are obtained from the mature plants in late summer and

fall. The stems can be dried and split, then the brittle inner part is broken away from the fibrous tissue just under the skin. The fibre, once cleaned, can be spun into a two-ply string, which is very strong and can be used for fish nets, duck nets and deer nets.

Stinging nettle is used to treat arthritis and has a range of other medicinal uses, including treating paralysis of the legs.

INDIAN HELLEBORE or
FALSE HELLEBORE (*Veratrum viride*)

LILIACEAE OR MELANTHIACEAE (LILY FAMILY)

NOTE: This plant is very poisonous, potentially deadly if used incorrectly.

HUL'Q'UMI'NUM' NAME: *Qwun'ulhp*

DESCRIPTION: A herbaceous perennial with an upright, unbranched flowering stem growing from a short, stout rhizome. The stems grow up to two metres tall, often with several clustered together. The stem leaves are distinctive, with prominently ribbed or pleated veins. They are oval to elliptic and up to 35 centimetres long, with smooth edges, smooth above with dense hair beneath. They diminish in size toward the upper part of the stem. The flowers are pale green or yellow-green with six similar spreading tepals (petals and sepals combined) and grow in a dense, branched and elongated terminal cluster, the lower branches drooping. The fruits are barrel-shaped, three-lobed capsules, each containing numerous seeds.

WHERE TO FIND: Moist streambanks, wet meadows and swamps and open forests from near sea level to (more commonly) montane areas and meadows; common throughout most of British Columbia, including suitable habitats within Quw'utsun territory.

CULTURAL KNOWLEDGE: This is a very important medicinal and spiritual plant for the Quw'utsun and W̱SÁNEĆ peoples, among others. It is a powerful plant and potentially very poisonous if used improperly. Luschiim noted that it used to grow at Goldstream but wasn't sure whether it can still be found there (April 18, 2016).

GIANT VETCH or BEACH VETCH

(*Vicia nigricans*, syn. *Vicia gigantea*)

FABACEAE (PEA FAMILY)

HUL'Q'UMI'NUM' NAME: *Tl'ikw'un* (Luschiim, April 18, 2016)

DESCRIPTION: A climbing perennial herb, growing up to two metres long, with hollow, angled stems. The leaves are pinnately compound, with 16 to 26 lance-shaped or oblong leaflets with well-developed tendrils. The flowers are small and pea-like, in dense, long-stalked clusters, reddish-purple or yellowish. The fruits are of a typical pea-pod type, three to five centimetres long, containing several small pea-like seeds. The pods and seeds are green at first, turning black at maturity.

WHERE TO FIND: Upper beaches, meadows, lagoons and forest edges in coastal British Columbia, including Quw'utsun beaches and estuaries.

CULTURAL KNOWLEDGE: Luschiim has seen this plant growing from Qualicum to Nelson's Point at Chemainus Bay to Malahat. He said, "You can eat them [the peas] ... You pick them when they're very young. When they're on the mature side ... I don't bother [with them]" (April 16, 2015). He said that by the time of our conversation (mid-April), it was almost too late;

the peas would be too old. (**NOTE:** One should not eat too many of these peas. They are potentially harmful in large quantities, or when they are mature.) This is also a spiritual plant, used for bathing, strengthening and protection (Luschiim, April 16, 2015, and April 18, 2016). The name *tl'ikw'un* is also used by some for beach pea, purple pea and garden pea (see page 229).

DEATH CAMAS (*Zigadenus venenosus*)

LILIACEAE OR MELANTHIACEAE (LILY FAMILY)

NOTE: This plant is very poisonous, potentially deadly if eaten. The bulbs resemble those of edible blue camas, so particular care is needed not to confuse them.

HUL'Q'UMI'NUM' NAME: Luschiim described this plant as *to' p'uq' tu sp'eq'um-s st'e 'ukw' speenhw' 'i qul-lhulh tth'ux-tun* (literally "like *speenhw'* but bad")

DESCRIPTION: A perennial herb growing from a deep, egg-shaped bulb up to five centimetres long, with upright flowering stems growing to 50 centimetres or taller. The leaves are grass-like and pointed; the basal leaves are up to 40 centimetres long, V-shaped in cross section and rough-margined. The flowers are creamy-white and crowded in a compact, pointed terminal cluster. The flowering stems elongate with maturity, and the fruiting capsules are three-parted, containing numerous brown seeds.

WHERE TO FIND: Meadows, grassy or rocky bluffs, forest edges and open woods from sea level to montane areas. These plants are common across southern British Columbia,

including in Quw'utsun territory; sometimes growing together with edible camas. The two are easily distinguished during flowering, but harder to identify when fruiting or in the fall and early spring.

CULTURAL KNOWLEDGE: Luschiim noted the importance of distinguishing between this plant and the edible, blue-flowered camas, and worries about potential poisoning. He said that formerly, the people would go and dig death camas plants out of edible camas areas to avoid the possibility of harvesting the poisonous one by mistake, since it's hard to tell the difference between the bulbs of **speenhw'** and death camas. He said that several people went to Mount Prevost (Swuq'us) and Mount Tzouhalem (Shquw'utsun) to dig the two species and check: the bulbs are a different shape and the stems and leaves are also different, but it is still important to weed them out because some people find it hard to distinguish between the bulbs. He said that death camas is a very valuable medicine when taken in small quantities, but it must be used with great care (April 18, 2016).

Bulbs of death camas (left) and common camas (right)

SOURCES

Interviews with Luschiim

Conducted by Nancy Turner (NT) and colleagues

December 13, 1999

March 28, 2005 (travel to Mount Prevost with NT and Greg Sam)

April 2005 (at Hul'q'umi'num Treaty Office)

May 2, 2005

September 24, 2010 (at Arrowsmith)

October 13, 2010

December 7, 2010

December 16, 2010 (with NT and Trevor Lantz)

January 23, 2011 (with NT and Kate Proctor)

June 24, 2011

November 2014

April 16, 2015

April 18, 2016 (with NT and Pamela Spalding)

May 25, 2017 (with NT and Genevieve Singleton)

February 14, 2019

References Cited

"Cowichan Elder Luschiim receives Honorary Doctorate of Letters from Malaspina University-College." June 18, 2007. *Vancouver Island University News.* https://news.viu.ca/cowichan-elder-luschiim-receives-honorary-doctorate-letters-malaspina-university-college/ (accessed May 1, 2016)

Cowichan Tribes Elders' Advisory Committee. 2007. *Quw'utsun: Hul'q'umi'num' Category Dictionary.* Quw'utsun Syuw'entst Lelum: Duncan, BC.

Cowichan Tribes. 2013. *Hul'q'umi'num' to English Dictionary.* Revised by Donna Gerdts, from previous versions prepared by Tom Hukari, Ruby Peter and Hul'q'umi'num'-speaking Elders. Duncan, BC. http://abed.sd79.bc.ca/wp-content/uploads/2013/09/Hulquminum_to_Eng2.pdf (accessed February 4, 2021)

Elmendorf, W.W., and W.P. Suttles. 1960. "Pattern and Change in Halkomelem Salish Dialects." *Anthropological Linguistics* 2(7): 22–23.

"Hul'q'umi'num': About Our Language." 2000. *FirstVoices.* https://www.firstvoices.com/explore/FV/sections/Data/Coast%20Salish/Halkomelem/HUL'Q'UMI'NUM'/learn (accessed May 3, 2016)

Kuipers, Aert H. 2002. *Salish Etymological Dictionary*. Linguistics Laboratory, University of Montana: Missoula, MT.

Rozen, David L. 1985. "Place-Names of the Island Halkomelem Indian People." MA thesis, Department of Anthropology and Sociology. University of British Columbia: Vancouver.

Turner, Nancy J., Leslie M.J. Gottesfeld, Harriet V. Kuhnlein, and Adolf Ceska. 1992. "Edible Wood Fern Rootstocks of Western North America: Solving an Ethnobotanical Puzzle." *Journal of Ethnobiology*, 12(1): 1–34.

Turner, Nancy J., and Richard Hebda. 2012. *Saanich Ethnobotany: Culturally Important Plants of the W̱SÁNEĆ People*. Royal BC Museum: Victoria.

Wilson, Captain Charles. 1866. "Report on the Indian Tribes inhabiting the country in the vicinity of the 49th Parallel of North Latitude." *Transactions of the Ethnological Society of London*, Vol. IV. John Murray Albermarle Street, London, UK: 275–332.

APPENDIX 1
General Hul'q'umi'num' botanical names and terms pertaining to plant use

(See *Quw'utsun: Hul'q'umi'num' Category Dictionary*, pages 145–49, for more detail.)

bake by steaming in a pit: *tth'hwas*

bake in a traditional method, covered with hot coals: *'atha'qw*

barbecue: *qw'ulum*

bark (inner bark of cedar): *sluwi'*

bark (thick bark of cottonwood): *qwoonulhp*

bark (thick tree bark): *p'uli'*

bark, thin (e.g., arbutus): *kw'ulo'*

basket, clam: *kwi'kwlha'lus* (name for a particular type with forked twigs)

basket, watertight: *skw'a'wus*

berry or berries: *stth'oom'*

berries, small: *s-tth'itth'oom'*

bitter: *sexum*

branch: *sh-ts'ushtutsus* or *s-ts'ushtutsus*

branches or plant stems, to cut: *lhtth'een*

buds (bursting open): *puxtssum*

burl (on a tree): *sts'up-hwun's tu thqet* (the same word is used for "wart"; Luschiim, September 24, 2010)

burn the ground: *yuqwunpt*

burned (like burning hair): *mexum*

cone of coniferous tree: *p'isuts'*

cooked or ripe: *qw'ul*

digging stick: *sqelux* (this is also the name for the little "worm-looking thing" inside a butter clam; Luschiim, December 2010, April 16, 2015)

edible cambium (and associated tissues): *sxa'muth-us* (see *sxe'muth*, "running sap"; Luschiim, December 2010)

emetic (substance that induces vomiting): *shhwuy'utun'um'*

flower: *sp'eq'um* (plural: *sp'e'luq'um*)

fruit: *stth'oom'* (berries); *shaalum* (fruit, general, including berries)

fungus (bracket or echo fungi): *tuw'tuw'eluqup*

fungus, multi-layered, unidentified: *hwuhwa'us-uhw*

garden: *qw'ext* (any kind of garden, including a clam garden)

grass or hay, general: *saxwul*

ground, lightly attached to: *shquli'qep'nuts* (see also rattlesnake plantain)

ground getting sour (infertile): *sasiyum'thut*

grow: *ts'iisum*

grow or cultivate something: *ts'uts'siim't*

grown over (e.g., with weeds): *ts'siima'*

juice (berry or fruit): *shqa'ul'uqw*

laxative: *shqw'ul-tun* or *shqw'uw'i'lum'*

leaf: *sts'alha'* (plural: *sts'a'lulha'*)

log: *qwlhey'*

moss (now used generically): *q'uts'i'*

mould: *papuqw*

needle(s) of a tree: *tth'e'lumutth'*

overgrown: *ts'siima'*

peeled: *kw'ulutth'* (like a peeled orange)

pick (e.g., fruit, berries): *lhumts'els*

pitch: *chumux*

plant or tree, suffix: *-alhp*. Luschiim recalled that there was a big discussion among the Elders back in the '60s and '70s about this suffix and what it meant. "Some of them decided it means a tree … many of our trees have that *-alhp* on there. So they all agreed, 'Yes, that's right,' and somebody else spoke up: How about this, then? How come [some other plants have that]?' And how about this one: *ququn'alhp*. That's your prince's pine [*Chimaphila umbellata*] … They're not trees, so further discussion—probably two or three years [later] —say, 'Well, it seems to mean something [else].' Medicines, many of them that are used for medicines have *-alhp* on the end, so we don't know … Trying to take apart our words and in trying to figure out what each part means" (April 16, 2015).

plant it; plant, sow or bury something (e.g., seeds, plants): *punut*

planted, sown or buried: *spu-pin'*

planted, sown or transplanted thing (e.g., trees, seeds, oysters transplanted from another area): *spun'um*

planting, sowing, burying or spawning: *pu-pun'um'*

poles: *tse'lumun* (Luschiim, December 2010)

pound (verb; e.g., to pound bark or a nail): *tth'asut*

pruning something: *t'qw'eent*

pull something out, uproot: *qw'umut*

pulling out plants, weeding: *qw'um'tsaan's*

root: *kwumluhw* (plural: *kwukwimluhw*)

root (small): *kwikwum'luhw*

sap (running on a tree): *sxe'muth*

seed(s) (e.g., grass seeds, alder cones, cedar and hemlock seeds): *tum'ekw'qun'*

seed(s) of fruit (e.g., peach or plum seed): *shtth'u-m'iwun*

shoots just emerging: *pulkween*

shoots of plants (e.g., salmonberry or blackberry shoots): *the'thqi'*

slivers in Douglas-fir bark: *sts'itsum'*

snag (standing dead tree): *st'epi'*

someone strips off bark: *lhqw'e'um*

strip off (e.g., bark from cedar tree or boards from a wall): *lhqw'at*

stump and root wad: *s-ulnuts*

transplant (e.g., young trees): *hwteyqnuts-t*

transplanting something: *hwteti'qnuts-t* or *hwtey'qnuts-t*

tree: *thqet*

tree (small): *thi'thqut*

trees or forest: *thuthiqut*

trunk (bole): *qwlhey's tu thqet*

tumpline: *tsum'utun* (also used for the strap-like part of a clam, and now used for "backpack")

weed (the garden, verb): *qw'um'tsaans* "to go and weed [the garden]"

weeds (noun): *squl'wey'*

wood: *syalh*

wood chips created by axe or chisel: *t'um'mun*

APPENDIX 2
Plants for which we found minimal information

———————

***Achlys triphylla* (vanilla-leaf).** Luschiim didn't know of any particular use for this plant (April 18, 2016).

***Adiantum aleuticum,* syn. *A. pedatum* (maidenhair fern).** Luschiim knew of no name for this delicate, black-stemmed fern, although he said it might be one whose stems were used in basket decoration. He noted that it grows right in the spray beside waterfalls and along creeks (April 16, 2015).

***Anaphalis margaritacea* (pearly everlasting).** Luschiim said that this plant had a name similar to that of kinnikinnick (*Arctostaphylos uva-ursi*), creeping snowberry (*Symphoricarpos occidentalis*) and pearly everlasting (*Anaphalis margaritacea*)—something like *tl'íkw'un'* (September 24, 2010, and April 16, 2015).

***Arctium minus* (common burdock).** The Hul'q'umi'num' names for this plant are *sexum,* "bitter," or *xuxumels,* "hawk" (because of the way hawks swoop down with their talons, which resemble the hooked bracts on burdock fruits). Luschiim knew of no use for burdock (June 24, 2011).

***Arctostaphylos uva-ursi* (kinnikinnick).** The Hul'q'umi'num' name of this low-growing evergreen shrub was not recalled by Luschiim, but he said it was similar to that of pearly everlasting (*Anaphalis margaritacea*) and creeping snowberry (*Symphoricarpos occidentalis*)—something like *tl'íkw'un'* (April 16, 2015). The leaves were dried and smoked by some people, but Luschiim had not heard of this use.

***Aruncus dioicus* (goatsbeard).** Luschiim recognized this plant but did not know of a name for it.

***Athyrium filix-femina* (lady fern).** Luschiim had heard of a name similar to *luq'luq'-'ey'* (a name suggested for the fiddleheads of this fern), but wasn't sure what it referred to (December 7, 2010).

***Betula papyrifera* (paper birch).** Elmendorf and Suttles (1960) gave the name *sukw'umuy'* for this tree. Luschiim says this is the Lummi name for birch (May 2017).

***Blechnum spicant* (deer fern).** Luschiim does not know of a name for this fern. He recognized the fern and said that it was not called by the same name as sword fern (April 16, 2015).

***Brassica rapa* (field mustard) and other weedy mustards.** Luschiim called wild mustard *sh hwiiliwe'een,* named after turnips, *sqwilíw.* He commented, "We had our own turnips out on the Gulf Islands." The mustard was said to come with the newcomers' oats, barley and other field crops (April 18, 2016; also discussed September 24, 2010).

***Claytonia sibirica* (Siberian miner's-lettuce).** Luschiim did not know of a name for this plant but said that it was supposed to be good eating.

***Clinopodium douglasii* (yerba buena).** The Hul'q'umi'num' name for this fragrant creeping vine in the mint family (Lamiaceae) is **stul-'á'tuxw** (pertaining to resting on the ground—this is a diminutive form; see **statuxw**, "codfish resting on the bottom"). It was, and is, used to make a beverage tea (April 18, 2016).

***Cornus canadensis* (bunchberry, or dwarf dogwood) and related species.** Luschiim recognized this plant but knew of no Hul'q'umi'num' name for it (September 24, 2010, and June 24, 2011). He suggested that a new name could be derived from the flowering dogwood tree: "All names started somewhere." The name **kwi'-kwitxulhp**, meaning "small dogwood," would be appropriate. He said the red berries could be used as a survival food, but that "it's really **ts'ewum'** ("no taste, bland"; June 24, 2011).

***Cytisus scoparius* (Scotch broom).** Luschiim said, "So, we attached a name to it, and we use the name "broom." Broom was your **'uxwtun**. [So the plant name became] **'uxwtunulhp**" (January 23, 2011).

***Dicentra formosa* (wild bleedingheart).** This is called **sp'eq'ums tu 'ulhqi**, "flower of the snake," but as Luschiim noted, this is not necessarily a negative name, since "some people get along well with snakes" (June 24, 2011).

***Dryopteris expansa* (spiny wood fern).** Luschiim did not know of a name for this fern and is not certain about the Hul'q'umi'num' names recorded elsewhere (April 16, 2015). (These other sources give the name as **luq'luq'-'ey'**, or **ntlk-klay**, but these words may refer to lady fern or other fern species.) This species of fern is named in many languages throughout northwestern North America, and the fern's rootstocks, which are said to resemble a pineapple or cluster of little bananas, were formerly pit-cooked and eaten, and also served as an emergency food in the wintertime.

***Fallopia japonica* (Japanese knotweed).** Luschiim said there was already lots of Japanese knotweed along the Cowichan River when he was a boy of around 10 years old; they figured it was from somebody's garden upriver.

***Maianthemum dilatatum* (wild lily-of-the-valley).** Luschiim did not know of a name for this plant, but said that people sometimes confused it with wild ginger because they both have heart-shaped leaves.

***Maianthemum racemosum* subsp. *amplexicaule* (false Solomon's-seal).** Luschiim did not know of a name for this plant (April 18, 2016).

***Marah oregana* (manroot, or wild cucumber).** The fruits resemble small cucumbers, but they are poisonous to eat. Luschiim possibly recalled the old people, long ago, talking about this plant (April 18, 2016).

***Menziesia ferruginea* (false azalea).** Luschiim recognized this shrub but didn't know of any name for it (April 16, 2015).

***Potentilla egedii* (Pacific silverweed).** Luschiim had heard that people eat the roots of this plant but did not know of a name for it.

***Prosartes hookeri* (fairybells).** Luschiim did not know of a name for this plant, which grows in moist places under trees. He said, "That one … could be a name, but I didn't get it. Its name would be referring to the two [flowers] … when I asked about it, that's what I was told" (June 24, 2011).

Ranunculus uncinatus, R. occidentalis, R. repens **(woodland buttercup, western buttercup, creeping buttercup) and others.** Luschiim did not know of a name for these flowers. He was told that creeping buttercup was introduced (April 18, 2016).

Rubus chamaemorus **(cloudberry).** Luschiim recalled a berry fitting the description of cloudberry (low, orange berry, like a salmonberry). He called it **s-tth'oom-s tu spulqwiitth'** ("berry of the ghost"), and said that you can eat them, but not too many or they would be harmful (May 25, 2017).

Rumex acetosella **(sheep sorrel, or sourgrass).** Luschiim did not know of a name for this plant but said that his cousins used to eat it (April 18, 2016).

Stachys chamissonis **var.** *cooleyae* **(hedge nettle).** Luschiim did not know of a name for this plant but said it is known as elk's food (April 18, 2016).

Symphyotrichum eatonii **(Eaton's aster).** Luschiim recognized a photo of aster flowers; he said they grew up in the mountains, but he didn't know of any name for them (June 24, 2011).

Tanacetum vulgare **(common tansy).** Luschiim did not know of a name or use for this plant. Others have called it **shew-qéen**, the same as the name for wild carrot or Queen Anne's lace (*Daucus carota*).

Taraxacum officinale **(common dandelion).** Luschiim did not know of a name for common dandelion, but said that the leaves were edible and that the white juices (latex) from the flower stem are good to treat warts, as are the roots; they are tied over the wart (April 18, 2016)

Trillium ovatum **(western trillium).** Luschiim commented that this is a very spiritual plant for the Quw'utsun (April 18, 2016).

Viola glabella **(yellow violet).** Some people told Luschiim that the Quw'utsun used to eat yellow violet, but this was no one in his immediate family; he did not know of any name for this plant (April 18, 2016).

Zostera marina **(eelgrass).** Eelgrass is possibly called **chulum**, similar to other Coast Salish names for this plant (April 18, 2016), but Luschiim had not heard of eating the rhizomes.

APPENDIX 3
Names of introduced garden plants and plant products

———————

apples: *'apuls* (borrowed from English)

beans: *piins* (borrowed from English)

blackberry, blackberries: *sqw'iil'muhw* (also used for native trailing blackberry)

bread: *suplil*

chewing gum (traditionally from a tree): *kw'i'hw*

coffee: *kafi* (borrowed from English)

corn: *na'tui*, or *kan* (borrowed from English)

flour: *spukw'*

gooseberry: *t'em'hw* (originally used for coastal black gooseberry)

grapes: *klips* (borrowed from English)

hops (*Humulus lupulus*): *heps* (borrowed from English). Some Quw'utsun people used to pick hops in the Agassiz area. Luschiim's mother, Violet Charlie, was there when he was small, around 1943 or 1944. He learned some of the bone game songs there (June 24, 2011).

onions: *'unyuns* (borrowed from English)

oranges: *'alunchus* (borrowed from English)

peas: *tl'ikw'un* (also used for native beach pea and giant vetch)

potato: *sqewth* (originally for wapato, or arrowleaf swamp potato)

pumpkin: *pumkun* (borrowed from English)

strawberries: *stsi'yu* (originally pertaining to wild blueleaf and seaside strawberries)

Raphia (*Raphia farinifera*): possibly named *sqwáthen*, but Luschiim doesn't know this name

turnips (*Brassica rapa*): *shhw'ileewe'* (also used for wild mustards; Luschiim, September 24, 2010).

vegetables, general: *s-ulhtunulhp*

INDEX: ENGLISH AND SCIENTIFIC PLANT NAMES

cow-parsnip, 227–28
crabapple, Pacific, 102–3
cranberry
    bog cranberry, 184–85
    highbush cranberry, 187
Crataegus douglasii, 130–31
currant
    red-flowering currant, 157
    stink currant, 151–52
cutleaf blackberry, 166–67

death camas, 257–58
deltoid balsamroot, 198–99
desert-parsley, barestem, 232–33
devil's-club, 145
dock
    bitter dock, 244–45
    western dock, 244–45
    yellow dock, 244–45
dogbane
    hemp dogbane, 193–94
    spreading dogbane, 193–94
dogwood
    Pacific flowering dogwood,
        98–99
    red-osier dogwood, 126–27
domesticated plum, 109
Douglas-fir, 60–67
Douglas maple, 87–88
dull-leaved Oregon-grape, 143
dwarf wild rose, 158

edible thistle, 211–12
elderberry
    blue elderberry, 168–69
    red elderberry, 170
Epilobium angustifolium, 209
Equisetum arvense, 33–34
Equisetum hyemale, 32
Equisetum telmateia, 33–34
Erythronium grandiflorum, 215–16
evergreen blackberry, 166–67
evergreen huckleberry, 183
Evernia prunastri, 14–17

false box, 146
false hellebore, 255
false ladyslipper, 200

fern
    bracken fern, 40–41
    goldenback fern, 35
    licorice fern, 36–37
    sword fern, 38–39
field mint, 236–37
fir
    amabilis fir, 44–48
    grand fir, 44–48
    silver fir, 44–48
    subalpine fir, 44–48
    See also Douglas-fir, 60–67
fireweed, 209
Fomitopsis pinicola, 20–21
foxtail barley, 225–26
Fragaria chiloensis, 217–18
Fragaria vesca, 217–18
Fragaria virginiana, 217–18
Fritillaria affinis, 219–20
Fritillaria camschatcensis, 219–20
Fucus distichus, 4
fungi
    bracket fungi, 20–21
    mushrooms, 22–23
    orange jelly fungus, 24–25
    shelf fungi, 20–21
    tree fungi, 20–21
    witch's butter, 24–25

Galium triflorum, 221
Ganoderma applanatum, 20–21
garden onion, 191–92
Garry oak, 114–17
Gaultheria shallon, 132–34
Geum macrophyllum, 222
giant horsetail, 33–34
giant vetch, 256
ginger, wild, 196–97
glasswort, American, 247–48
goldenback fern, 35
Goodyera oblongifolia, 223
gooseberry
    coastal black gooseberry,
        153–54
    gummy gooseberry, 155–56
    sticky gooseberry, 155–56
grand fir, 44–48

grasses and grass-like plants,
        225–26
    American dunegrass, 225–26
    beach wildrye, 225–26
    common rush, 225–26
    foxtail barley, 225–26
    speargrass, 225–26
great camas, 201–7
Grindelia stricta, 224
gummy gooseberry, 155–56
gumweed, Oregon, 224

hairy manzanita, 125
hardhack, 173–74
hawthorn, black, 130–31
hazelnut, 128–29
hellebore
    false hellebore, 255
    Indian hellebore, 255
hemlock
    mountain hemlock, 82–83
    western hemlock, 82–83
hemlock-parsley, Pacific, 213–14
hemp dogbane, 193–94
Heracleum maximum, 227–28
highbush cranberry, 187
Himalayan blackberry, 166–67
Holodiscus discolor, 135–37
honeysuckle
    orange honeysuckle, 139
    twinflower honeysuckle, 140
hooker's onion, 191–92
Hordeum jubatum, 225–26
horsetail
    branchless horsetail, 32
    common horsetail, 33–34
    giant horsetail, 33–34
huckleberry, 177
    black mountain huckleberry,
        182
    evergreen huckleberry, 183
    red huckleberry, 186
Hypogymnia spp., 14–17

Indian consumption plant,
        232–33
Indian hellebore, 255
Indian-pipe, 238–39

INDEX: HUL'Q'UMI'NUM' PLANT NAMES

ABOUT THE AUTHORS

Luschiim Arvid Charlie was born in Quamichan, one of the Cowichan Villages, in 1942 and has lived in the Duncan area all of his life. From the age of three, he began learning about plants and their various uses from the Elders in his family. Since then, he has made it a personal priority to gather knowledge about the natural environment. In 2007, he received an Honorary Doctorate of Letters at Malaspina University-College in recognition of his extensive contributions to the teaching of Coast Salish culture and traditions in a wide range of contexts, as well as his commitment to the protection of the environment and preservation of the Hul'q'umi'num' language.

Nancy J. Turner is internationally known for her work in ethnobotany, the study of plants and cultures. She is a Distinguished Professor Emerita in the School of Environmental Studies at the University of Victoria and holds honorary degrees from Vancouver Island University, University of Northern British Columbia, Simon Fraser University and the University of British Columbia. Turner has published over 30 books and over 150 scholarly papers and popular articles. For many years Nancy Turner has worked closely with Indigenous Elders—her teachers, collaborators and friends—to record their knowledge and understanding of plants, ecology and traditional stewardship practices. In collaboration with many First Nations, she has helped develop and support programs for retaining, enhancing and promoting the rich heritage of traditional botanical knowledge within communities. She lives in Nanaimo, BC.